T0303887

Biological Computation

CHAPMAN & HALL/CRC
Mathematical and Computational Biology Series

Aims and scope:

This series aims to capture new developments and summarize what is known over the entire spectrum of mathematical and computational biology and medicine. It seeks to encourage the integration of mathematical, statistical, and computational methods into biology by publishing a broad range of textbooks, reference works, and handbooks. The titles included in the series are meant to appeal to students, researchers, and professionals in the mathematical, statistical and computational sciences, fundamental biology and bioengineering, as well as interdisciplinary researchers involved in the field. The inclusion of concrete examples and applications, and programming techniques and examples, is highly encouraged.

Series Editors

N. F. Britton
Department of Mathematical Sciences
University of Bath

Xihong Lin
Department of Biostatistics
Harvard University

Hershel M. Safer

Maria Victoria Schneider
European Bioinformatics Institute

Mona Singh
Department of Computer Science
Princeton University

Anna Tramontano
Department of Biochemical Sciences
University of Rome La Sapienza

Proposals for the series should be submitted to one of the series editors above or directly to:
CRC Press, Taylor & Francis Group
4th, Floor, Albert House
1-4 Singer Street
London EC2A 4BQ
UK

Published Titles

Chapman & Hall/CRC Mathematical and Computational Biology Series

Biological Computation

Ehud Lamm

Ron Unger

CRC Press
Taylor & Francis Group
Boca Raton London New York

CRC Press is an imprint of the
Taylor & Francis Group an **informa** business
A CHAPMAN & HALL BOOK

Chapman & Hall/CRC
Taylor & Francis Group
6000 Broken Sound Parkway NW, Suite 300
Boca Raton, FL 33487-2742

© 2011 by Taylor and Francis Group, LLC
Chapman & Hall/CRC is an imprint of Taylor & Francis Group, an Informa business

No claim to original U.S. Government works

Printed in the United States of America on acid-free paper
10 9 8 7 6 5 4 3 2 1

International Standard Book Number: 978-1-4200-8795-6 (Hardback)

Library of Congress Cataloging-in-Publication Data

Lamm, Ehud.
   Biological computation / authors, Ehud Lamm, Ron Unger.
      p. cm.
   Includes bibliographical references and index.
   ISBN 978-1-4200-8795-6 (hardcover : alk. paper)
   1. Biocomputers. I. Unger, Ron. II. Title.

QA76.884.L36 2010
006.3'2--dc22
                                               2010028310

Visit the Taylor & Francis Web site at
http://www.taylorandfrancis.com

and the CRC Press Web site at
http://www.crcpress.com

Table of Contents

Preface

THE SPIRIT OF THIS BOOK

It is often said that biology is going to be the science of the 21st century as physics was the science of the 20th. Fascinating discoveries about the living world around us, as well as about our own bodies, are brought about daily by molecular biology, neuroscience, and other biological disciplines. In addition, biological understanding, whether on the molecular scale or on the ecological level, is fast becoming the foundation of new engineering disciplines, such as nanotechnology and bioengineering, which have the potential to fundamentally change the way we live.

Computers, and computer science ideas and techniques, are of course an important part of all these scientific and engineering activities. Computer science and its concepts and methods are not only a servant of biological research but also provide mental models used by a new generation of biologists, who often refer to themselves as systems biologists, in thinking about the living world. Ideas and approaches, however, travel in both directions: reflecting on biological ideas has inspired a wide range of computer science questions and has led to the development of important new techniques for solving hard computational problems. The result might be called **biological computation** (or biologically inspired computing) and is the subject of this book.

This book is written from the perspective of computer scientists who are fascinated with biology. A large part of the excitement and fun of bio-inspired computing, at least for us, is learning the amazing and quirky details discovered by biologists. Among the biological stories that have informed computer science you will find discoveries about how bacteria communicate, how ants organize their nests, and the way the immune system learns to recognize pathogens before actually encountering them. All these, and more, are discussed in the chapters to come, along with the computational techniques they led to. We hope the book manages to convey the sense of wonder and fun that we feel about the field. It goes without saying that it is impossible to go into all details of such varied phenomena, and we concentrate on the aspects of the biological phenomena most closely related to the computational approaches we discuss.

THE CONTENT OF THE BOOK

The term *biological computation* encompasses quite a few approaches. In this book we focus on the most fundamental and important ideas, and on the classic works in each of the subjects we discuss, in an attempt to give a unified overview of computer science ideas inspired by biology. The four major topics we focus on are **cellular automata, evolutionary computation, neural networks,** and **molecular computation**. Each of these topics is the subject of a chapter that begins by exploring the biological background and then moves on to describe the computational techniques, followed by examples of applications and a discussion of possible variants of the basic techniques introduced in the chapter. Each chapter also includes exercises and solutions. Exercises with solutions are marked with bold numbers. Important ideas and techniques are presented through the example applications and exercises. In addition to the chapters discussing these techniques, Chapter 1 provides a general biological background, and Chapter 6 concludes the book by introducing, more briefly, some of the new topics that are emerging within the field.

We made a special effort to make our explanation of molecular computation accessible to readers who lack a background in molecular biology, without sacrificing the details. In contrast to the other techniques we discuss in the book that can be immediately used by programmers to attack computational challenges, molecular computation is still mostly in its infancy and requires equipment that can be found only in professional laboratories. We feel, however, that thinking about the computational power of molecular events is enlightening, and we predict that computer scientists will enjoy the puzzle-like challenge of trying to represent computational problems as sets of interacting molecules. With the possibility of biological hacking and "Do It Yourself Biology" just around the corner, the use of these techniques may become more widespread than can presently be imagined.

The topics we focus on, with the exception of molecular computing, are already the subject of several good textbooks, which can be found listed in the Recommendations for Additional Reading section of Chapter 6. Most of these books, however, are dedicated to only one of the subjects we discuss or are extremely detailed reference books. Our goal was to present to you, the reader, an overview of the terrain, allowing you to then focus your attention on the techniques that are most relevant for you. Each of the approaches we cover exists in a multitude of variants and is covered by a large amount of theoretical work—it is very easy to get buried in the details. This book attempts to convey in an easily digestible form the gist of each of the major approaches in the field and to bring you to the point where you can produce a working implementation of each of the basic techniques or to effectively use one of the many existing implementations that can be found online. All the details can be easily found in the literature or by searching online once the basic ideas introduced here are understood.

The techniques we discuss reflect fundamental principles whose applicability goes beyond bio-inspired computing—for example, self-organization, redundancy, using noise, asynchronicity, nondeterminism, and other methods of parallelism and distributed computing. These ideas manifest themselves in other areas of computer science and software engineering, specifically in the development of very large-scale distributed systems, of the sort underlying cloud and grid computing. While these fields are not discussed here, we feel that getting acquainted with these fundamental ideas and playing with simple computational models that exhibit them, such as the ones presented throughout this book, can be rewarding.

FOR WHOM IS THIS BOOK INTENDED?

We wrote this book thinking primarily of readers with a computer science background and we assume no previous background in biology. For readers who feel they would benefit from a deeper understanding of the biological context we provide references to several recommended books in the Further Reading list in Chapter 1. This book is intended to be a gentle introduction to the field and should be suitable for self-study as well as for use in university courses. We assume the reader is familiar with basic computer science terminology and basic algebra and probability theory but provide detailed explanations of all derivations. There are programming exercises at the end of each chapter, but it is possible to follow the explanations and discussions without programming knowledge. We did not include many formal proofs, but throughout the chapters and exercises we give easy-to-follow examples of several important proof techniques. This should make the book accessible to readers with biological or medical backgrounds—those coming to the field of bio-inspired computing from biology rather than from computer science.

USING THIS BOOK AS A TEXTBOOK

While the book can be used for self-study, its main purpose is to serve as a textbook for a course on biological computation. Such a course can be given to advanced undergraduate or early graduate students in programs that combine biology and computer science (a double major in computer science and biology or special bioinformatics tracks). For such students a course on biological computation can complement a suite of courses in bioinformatics, algorithms for computational biology, and systems biology.

A course based on this book can also be given to students who major in computer science and for whom a course in biological computation can enrich the perspectives about alternative models of computation. The book contains in the first chapter and in each one of Chapters 2–5 an accessible biological introduction. Nevertheless, it is a good idea for these students to take a basic course in biology prior to a course based on this book or, as was done in Bar-Ilan University, to add lectures and teaching assistant (TA) sessions giving a "crash course" in biology.

The material covered in this book can be delivered in a semester (13–14 weeks) with weekly two-hour lectures and weekly TA sessions. Thus, we devote about three weeks to each one of the four main subjects we cover. From our experience, students gain a lot from homework and especially from the programming exercises, so we provide a good number of those. As the book was written as a textbook, we tried not to overwhelm the readers with footnotes and references.

When we designed and delivered this course in the last several years, no suitable textbook was available, and we felt its absence. Our course was well received by students, and we hope that this book will encourage and enable many teachers and universities to offer similar courses.

ACKNOWLEDGMENTS

Writing a book is a long and complicated process, and we could not have done it without the help and support of many individuals and institutions.

The idea to collaborate on a book came to us while we worked on the development of a course on bio-inspired computing for the Open University of Israel. The structure of this book is based on courses taught by Ron Unger at the Weizmann Institute of Science and for many years at

Bar-Ilan University, and the course developed by the two authors for the Open University of Israel. Our thanks go to these institutions.

Special thanks go to Edna Wigderson, who helped us transform a first draft written in Hebrew into the book before you. Not only did she translate our original material, but she also edited the material, pointed out inconsistencies and mistakes, and helped us improve the presentation immeasurably. Without her this book would never be.

We thank Assaf Massoud for the artwork resulting in the illuminating illustrations that accompany the text. Working with Assaf was a real pleasure. Both Edna and Assaf had to endure the difficulties of dealing with two authors who often disagreed; not only did both endure this with grace, but their prodding also helped the two authors converge.

We also wish to thank all those who read the manuscripts or parts of it, pointed out our mistakes, and made valuable suggestions. First and foremost we thank Yair Horesh, who was involved in the courses in Bar-Ilan and in the Open University and made important contributions to the manuscript. We would also want to recognize the assistance we received from Tania Gottlieb in the biological aspects of the book and Orly Noivirt-Brik, Yochai Gat, Nurit Zer-Kavod, Ariel Azia, Tirza Doniger, Inbal Yomtovian, Ari Yakir, and Ilana Lebenthal for their valuable comments on the manuscript.

As is always the case, the responsibility for whatever errors remain is ours alone.

On a personal note, Ron wishes to thank David Harel, Joel L. Sussman, and John Moult, mentors and friends who helped him in his transition from computer scientist to computational biologist, and his colleagues at Bar-Ilan University and especially Elisha Haas and Shula Michaeli. Special thanks go to my family—my parents Zipora and Ozer; my sister Vered, who was so helpful in preparing the book; and my wonderful children Amir, Ayelet, Hilla, and Inbal. I want to express special thanks to my vibrant wife Tamar. Tami, without your support and love, this book project could not have been undertaken, let alone completed.

Ehud wishes to thank Eva Jablonka for intellectual stimulation and biological insight, in addition to friendship and moral support; his many colleagues in the computer science department at the Open University of Israel for encouragement and support; and Dror Bar-Nir and Sara Schwartz for invaluable discussions. I wish to thank my family for their love and support. I am especially grateful to my wife Ayelet whose love and friendship are the greatest gift of all.

CHAPTER 1

Introduction and Biological Background

1.1 BIOLOGICAL COMPUTATION

This book presents topics on the border between biology and computer science in an attempt to demonstrate how biological insights allow us to deal with complex computational problems and, conversely, how computer science insights enhance our understanding of biological processes and help to identify problems worthy of research.

Most of the book presents the topics from a computer science perspective and deals with new computational models and techniques based on ideas derived from biological research. Using these techniques, problems are solved in ways that differ from "classical" computer programming, in which programs can usually be described as a linear sequence of instructions. Chapter 2 deals with cellular automata, which are made up of many independently operating cells embedded on a grid, each of which can affect only its neighbors. This resembles a colony of simple organisms (e.g., bacteria) that can present amazingly complex behaviors or even the structure of simple multicellular creatures that contain many cells working together. Chapter 3 deals with evolutionary computation and demonstrates how to solve optimization and search problems by mimicking the natural evolutionary processes whereby organisms adapt to their environment. Chapter 4 presents models of neural networks that attempt to mimic the behavior of the brain. These systems are capable of learning and generalizing from examples. Chapter 5 deals with molecular computation, in which computational problems are solved by a set of interacting biological molecules. Finally, Chapter 6 presents brief descriptions of several other topics that are on the cusp between computer science and biology, for example, ideas drawn from animal behavior and from the operation of the immune system.

We will explore how to apply the models presented to a large variety of computational problems. In particular, some of the problems discussed are believed to be hard (e.g., we will discuss NP-complete problems, which are problems for which no efficient algorithm is believed to exist), but using ideas inspired by biology helps in finding practical solutions for many of their instances.

We can consider the use of biological insights to solve computational problems as a "translation" of biological phenomena into formal mathematical models. Obviously, we will not attempt an exact translation but will merely use certain aspects of the biological phenomena as inspiration for developing mathematical and computational methods. One could also do the converse—develop formal mathematical tools to analyze biological phenomena. This field is called **theoretical biology** and is beyond the scope of this book. The discussion of systems biology in Chapter 6 briefly notes how some of the theoretical ideas developed by computer science are being applied to the study of biological systems.

As we discuss these new computational models we will naturally focus on the differences between them and traditional computational models. Table 1.1 outlines a few major differences between the biologically inspired models and traditional, more conventional models. The table demonstrates that nonstandard computational models differ from conventional models in many significant aspects.

Note that, in addition to the biologically inspired computational models, other types of nonstandard computational models have drawn a lot of attention recently. An interesting example is **quantum computing**, which attempts to use physical properties described by quantum mechanics as a powerful computational mechanism. As we concentrate on biological ideas, these models will not be discussed in this book.

These new approaches to computation are both intellectually exciting and present new engineering approaches to solving complex problems. It almost seems as if the biologically inspired models were designed to deal with difficulties and limitations that arise when building complex computer systems. The engineering requirements arising from the attempts to build increasingly complex computer systems that have to be reliable

TABLE 1.1   Biologically Inspired Models versus Standard Models

	Conventional Computation	Biological Computation
Mode of operation	Mainly *sequential*, even though there exist parallel computers and it is possible to write parallel code that executes on standard hardware. Usually the number of parallel processes is very small.	Mainly *parallel*. Most biologically inspired models are massively parallel and are based on thousands of local interactions running in parallel.
Control	There is a centralized *global control* of the entire system.	Computation is the result of numerous local processes without a global control mechanism.
Programming	The programmer has to specify in detail the behavior of the system, by, for example, choosing appropriate data structures or algorithms.	In the models presented in Chapters 3 and 4 central aspects of the behavior of the system are developed by the system gradually and evolutionarily, without human intervention.
Modifiability and adaptability	Depends on the design of the system but usually requires re-programming.	The systems are able to adapt to a wide range of environmental changes and changes in the system itself, without external intervention.
Robustness and error tolerance	Requires special treatment and usually requires dedicated code or hardware. Usually systems are incapable of dealing with extensive or prolonged failure (e.g., a hardware failure), and such failures usually end in catastrophic behaviors (complete shutdown of the system).	The model often leads to inherent robustness. Some of the models are capable of independent gradual correction of widespread failures.
Requirements from components	In general, a reliable system needs reliable components, and the system is reliable as the weakest component it includes. Usually, nondeterminism cannot be tolerated.	It is possible to build a reliable and fast system using unreliable, slow and noisy components. Nondeterminism at various levels of the system can even contribute to reaching the system's functional goals.
Hardware	Electronic, usually silicon based.	Usually implemented on standard hardware. Chapter 5 describes using organic (carbon-based) materials for computations.

and efficient (therefore often parallel and distributed) and to make use of cheap components to reduce overall costs seem remarkably suited to the properties of nonconventional models. Moreover, throughout the book we will encounter insights gleaned from nonstandard models that can be integrated into more conventional computer systems.

It is important to understand that many other techniques are employed when developing complex computer systems to satisfy the aforementioned criteria. Large systems such as those used by Google® and Amazon® have to deal with huge datasets, to satisfy requests from all over the globe, and to be very reliable and highly available. To achieve these goals they deploy several large data centers, each of which houses a large number of clusters of computers. This necessitates synchronizing a large number of servers while, for example, minimizing latency, performing computations in parallel, distributing computations between machines and data centers, and automatically recovering from faults (e.g., using redundancy). Such computing and storage architectures are becoming more and more available to even smaller companies (e.g., Amazon® sells services on its computing "cloud"). The architectural complexity of the system can be hidden from the programmer so that only the infrastructure designers and implementers need to worry about these details and the programmer can concentrate on the details of the applications being developed.

In some instances we can find a similarity between the fundamental principles being used in the development of large distributed systems and the principles underlying biologically inspired computational models, even though the systems themselves differ significantly. The similar principles include self-organization, redundancy, use of noise, nondeterminism, and other methods of parallelism and distributed computing. We will demonstrate these general principles throughout the book.

1.2 THE INFLUENCE OF BIOLOGY ON MATHEMATICS—HISTORICAL EXAMPLES

Intellectual disciplines interact with each other in many different ways. Some of these interactions have a direct effect on fields of study, while others are more circumstantial. Examples of the direct interactions include using techniques developed in one field of study to analyze phenomena in another field or a research field splitting up into more specialized subfields. The indirect interactions might arise when new ways of thought affect multiple disciplines (e.g., with the rise of statistical thinking), when new research directions open up, or when researchers switch fields.

It would seem that biology cannot have much of an impact on mathematics (and later on computer science), as these disciplines are fundamentally so different. Mathematics is exact and deals with formal arguments and proofs. Consider, for example, Euclidean geometry developed by Euclid in the book *Elements* in the third century BCE. Ancient as this mathematical field is, its results are still valid (and will be valid eternally), as they are logically derived from the axioms Euclid laid down in his classical work. Biological research, on the other hand, evolves constantly. New facts are discovered at a great pace, and a large chunk of our current biological knowledge is derived from recent research. Moreover, biology deals with a vast array of phenomena, which we understand only partially and imprecisely, and this understanding also keeps changing and evolving. Despite these difficulties, mathematical tools are often needed for analyzing biological phenomena, and when no appropriate tools exist they need to be developed. This, in turn, leads to the emergence of new mathematical and computational fields. We present in this section some examples of the influence of biological research on mathematics in the past. The later chapters are indicative of current cross-influences and hint at future developments.

In the years 1827–1828 the Scottish botanist **Robert Brown** (1773–1858) published his findings about the motion of pollen in water. Using a microscope (an advanced technology for his time), he discovered that the pollen particles swirled about in a motion that could not be explained by the water movement. Initially, after observing pollen derived from different flowers, Brown believed that this motion is observable only in particles derived from living matter. This conclusion was in step with accepted biological theories of the time. However, further research established similar behavior in tiny particles of inanimate matter, such as glass. Further study of this "Brownian Motion" revealed its probabilistic nature. Today, it is well known that this phenomenon arises due to the impact generated by the random collisions of the observed particles with the many much lighter water molecules surrounding them (a water molecule's radius is roughly 1.4 Å (1 Ångström = 10^{-10} meters), while the radius of a pollen particle is of the order of 10 micrometers (1 micrometer = 10^{-6} meters)). The mathematical analysis of this phenomenon had a large impact on the development of probability theory (Albert Einstein was one of the contributors to this analysis). This theory is heavily used today to analyze a wide array of probabilistic processes that have nothing to do with biology or molecular motion.

The Brownian motion example is not unusual. There is a close link between the development of statistics and the study of probabilistic (or stochastic) processes and the pursuit of biological questions. One of the reasons for this link was the endeavor to collect, tabulate, and analyze human populations (e.g., conducting a census of a state's population).

Adolf Quételet (1796–1874) collected and analyzed height and weight data and was amazed to discover that their distribution was a normal distribution (a "bell curve"). Up to his time normal distributions were used only to explain measuring errors, mainly of astronomical phenomena. Some of Quetelet's discoveries were that the height of French army recruits was normally distributed, as were the chest measurements of Scottish soldiers. Quetelet denoted the center of these distributions by the term *the average man* ("l'homme moyen") and believed that one can study social phenomena by observing the differences in distributions among different groups of people (e.g., among different races).

Quetelet's ideas had a profound influence on **Francis Galton** (1822–1911), who was a cousin of Charles Darwin and a polymath. Among his other contributions to statistics, Galton pioneered the use of regression for finding a linear function best describing a set of data and the notion of correlation. He used these tools when investigating the heritability of properties such as height (i.e., the relation between the height of parents and their children) and the size of peas. Galton also dealt with the problem of the disappearance of certain family names over time. This research had influence on the study of *stochastic processes,* which in turn are used, among many other things, for researching the propagation of diseases (epidemiology).

As a last example we'll mention **Karl Pearson** (1857–1936), who was a student of Galton. He also dealt with biometrics (measuring biological properties). Among his many contributions is the **Pearson correlation coefficient**, which is used to describe the quality of the correlation between two random variables. Current medical research is based to a large extent on statistical tools, and its quality depends largely on the planning of clinical trials and a careful analysis of their results, using the techniques we mentioned among other tools. These tools form the basis for the modern concept of evidence-based medicine.

In 1948, **Norbert Wiener** (1894–1964) defined the term **cybernetics** to describe the study of control systems. A central element of regulatory and control systems is **feedback**, which occurs when two parts of a system interact in a bidirectional fashion so that they influence each other. For example, think of a system composed of components A and B, such that A influences B's behavior and B influences the behavior of A. The two major kinds of feedback are **positive feedback** and **negative feedback**. In a positive feedback loop the system responds to the perturbation by further changes in the same direction as the perturbation, whereas in negative feedback the system responds in the opposite direction and attempts to revert to the initial state. We will see that feedback mechanisms allow systems to self-organize and adapt to their environment.

Living organisms are capable of retaining their internal states in response to a wide range of perturbations in their environment. For instance, internal body temperatures in warm-blooded animals do not change with the temperature of the surroundings (as long as those changes are not too extreme). Similarly, athletes who train at high altitudes with lower oxygen concentration develop more red blood cells to maintain the amount of oxygen reaching their cells. This ability to maintain a steady internal state is called **homeostasis** (homeo = similar, stasis = standing still). Feedback loops play a central role in maintaining homeostasis.

Wiener defined cybernetics as the science dealing with control and communication in man-made and biological systems and was influenced by biological examples. Even though cybernetics is no longer considered an independent research field, its concepts and the basic problems it dealt with are still used in designing and analyzing dynamical systems.

Another area that posed major mathematical challenges is demographics, which is the study of populations (human and otherwise). Demographic tools are important for forecasting population sizes and for understanding the influence of habitat changes on populations. The mathematical description of demographic processes uses complex classes of differential equations. The need to study and solve these equations gave rise to various mathematical developments.

1.3 BIOLOGICAL INTRODUCTION

In this section we present a few topics in biology with the goal of providing the basic vocabulary needed for discussing biological phenomena. We will limit ourselves to discussing general background material necessary for understanding the following chapters, and more specific biological topics will be explained in later chapters.

Biology deals with living organisms. When encountering an object, we find it trivial to decide whether it is a living object, but are we capable of defining the difference between an animate and inanimate object? Can we explain the basis of the differences? The answer cannot be found at the physical level since living organisms are composed of the same building blocks as all other matter: atoms, which are made up of protons, electrons, and neutrons. The atoms are combined to create larger building blocks called molecules. For instance, a water molecule contains two hydrogen atoms and one oxygen atom. Living organisms contain carbon based molecules, known as organic molecules. **Organic molecules** are encountered almost exclusively in living objects and as products of living organisms. The main types of organic molecules are **proteins**, **carbohydrates** (sugars), **lipids** (fats), and **nucleic acids—ribonucleic acid (RNA)** and **deoxyribonucleic acid (DNA)**. These molecules play a major role in almost all processes occurring in any living organism.

The basic unit of living organisms is the **cell**. Organic molecules are created by cells, are used for cellular activities, and make up cells. Even simple single-cell organisms such as bacteria and yeast display a large spectrum of types and behaviors: for example, some survive better at high temperatures; some prefer colder environments; some require oxygen, and others do not; they have different nutrient requirements. More complex organisms made up of more than one cell are called **multicellular organisms**.

One of the first tools needed to understand and research living organisms is the ability to systematically group and categorize organisms. The classification system is based on observing similarities between different organisms. For instance, cats are more similar to each other than they are to dogs; cats, lions, and tigers present many similarities to each other and are different from dogs and wolves. Therefore, it is reasonable to impose a hierarchical structure on these animals: all cats belong to the "cat" group, which is a subset of the "feline" group that also contains tigers and lions, and so on. The biological disciplines of **systematics** and **taxonomy** deal with these classification problems.

A **species** is the lowest rung of the biological hierarchy. A species is commonly defined as a group of organisms capable of interbreeding and producing fertile offspring. Cats raised as pets ("house cats") all belong to one species, while all the feline species share various similar characteristics. For example, here are the main categories used to scientifically identify house cats:

Kingdom: Animalia (animals)
  Phylum: Chordata
    Sub-Phylum: Vertebrata (vertebrates)
      Class: Mammalia (mammals)
        Order: Carnivora (carnivores)
          Family: Felidae
            Genus: *Felis*
              Species: *Felis catus*

Tigers and lions, like the house cat, belong to the Felidae family but not to the *Felis* genus. Organisms that are more similar to the house cat such as the jungle cat (*Felis chaus*) do belong to the *Felis* genus.

Every species and every higher-order class contains organisms with similar characteristics, and researchers specializing in each group can recite a large body of scientific knowledge about the particular properties unique to each group. In our discussion here we will be mostly concerned with universal characteristics that apply to all organisms or at least to a large set of species. This chapter will describe some of these universal characteristics. In Chapter 4 we will discuss neural networks that are inspired by general principles observed in the nervous systems of highly developed organisms. Chapter 6 discusses computational models inspired by the immune system, which is found only in a subset of species, and by the behavior of social insects.

Similar characteristics may indicate either that different species evolved from a common ancestor or that they had to deal with similar environmental challenges. For example, bats and bees have wings used for flying, but their wings do not come from a common ancestor. The term **analogy** is used to describe organs or structures with an identical function, while a similarity due to shared common ancestry is called **homology**. Homologous structures may differ significantly in function and shape. For instance, the limbs of whales, bats, and humans are homologous. Another interesting example is the homology between the inner ear bone structure in mammals and the jaw bones of fish. The source of the universal characteristics of all living organisms with which we will deal in this book is very early in the evolution of life.

It is now accepted that each living organism belongs to one of three **domains**. A basic distinction is between organisms with cells that have a nucleus, called **eukaryotes**, and those that do not. A **nucleus** is a component in the cell that contains most of the cell's genetic material, or

genome (we will discuss the cell structure in greater detail in the next section.) This distinction has been known for a long while as it is often possible to observe the nucleus using a simple light microscope. With the advance of technology and molecular tools it became evident that two types of cells do not have nuclei—*Bacteria* and *Archaea*. The evolutionary distance between *Bacteria* and *Archaea* is large and it is wrong to view them as two subsets of the family of all nucleus-less cells. Rather than that, they are two separate classes at the same hierarchical level: **Bacteria** and **Archaea** are the domains of the **prokaryotes** (nucleus-less cells), whereas all organisms whose cells have a nucleus belong to the **Eukarya** domain. Multicellular organisms, including the species of Animalia, belong to *Eukarya*, but there are also eukaryotic single-cell organisms.

You might have wondered where viruses fall into according to this classification. Remember that we discussed the classification of organisms made of cells, but viruses are not cells and do not have all the mechanisms that allow a cell to use energy, to manufacture proteins, and to reproduce. Lacking these capabilities, it is debatable whether viruses should be considered living organisms. Viruses contain only genetic material and reproduce by penetrating a cell and reprogramming it to execute the instructions contained in the genetic material of the virus. As such, viruses may be considered the ultimate parasites.

Single-cell organisms (whether they are eukaryotes or lack a nucleus) need not live in isolation. Bacteria form colonies that may be millions strong. The bacteria in a colony not only live in proximity but also assist each other by creating organized sheets of cells, which make it easier for the bacteria to attach to the surface on which they live. For instance, dental plaque is made up of bacteria that form a biofilm. Bacteria colonies can even behave in an organized manner as can be seen when an obstacle is put in the way of bacteria in a petri dish and the colony circumvents it (we discuss this topic again in Chapter 2, where you will find photographs showing this remarkable process). ***Dictyostelium discoideum*** is a particularly interesting example of cooperation among single-cell organisms. It is composed of soil amoebae that usually live and reproduce independently. When their living conditions deteriorate—for example, when there is a food shortage—they become organized into a complex multicellular structure that can contain up to 100,000 cells and is surrounded by an extracellular skeleton. This slug-like structure can react to temperature changes and move as a single entity. In the next chapter we will discuss a computational model that can be compared to a colony of simple organisms and will show how systems exhibiting complex collective behavior can be created within the colony.

A fundamental difference between a multicellular organism and a colony of single-cell organisms is the **differentiation** of the cells into different cell types, each of which has unique properties and well-defined roles in the system. Differentiation is mainly a unidirectional process, whereby differentiated cells cannot change into a different kind of cell and can survive only within the multicellular organism. For example, a neuron cannot turn into a muscle cell. Only the genetic material existing in specialized cells (the **gametes** or **germ cells**) is used in the reproduction of the multicellular organism, while all other cells forego independent reproduction. Some cells (e.g., the human red blood cells) do not even contain a nucleus and genetic material. **Stem cells** exist in many tissues and in contrast with other cells have the capability to reproduce and differentiate into different types of cells and therefore have obvious medical potential. They have been heavily researched in recent years.

When observing the organization and division of labor within multicellular organisms, it is evident that there are many levels of organization, from the single cell up to the whole organism. Cells are organized in **tissues**, which have specific functions. For instance, muscle cells can contract and thereby allow the organism to convert energy into motion, allowing it to operate, for example, the respiratory muscles and the heart. Neurons are used for internal communication and also make up the central nervous system (brain) in organisms that have one. Blood cells (notice that blood is considered a tissue) are used to transport oxygen and nutrients to the cells and to support the immune system and also have other functions. Fat cells are used to store energy for times of need.

Cells belonging to different tissues are organized in **organs**, which have a specific function in the organism such as the heart, lungs, or brain. Organs that interact with each other closely to perform a certain function essential for the organism's survival are called a **system**, such as the respiratory system, the nervous system, the immune system, and the reproduction system. Finally, the systems build up the whole multicellular organism. The origin of such a modular organization is a central topic in our attempts to understand life (we will see in the next section that even a single cell has a modular structure with different components responsible for fulfilling specific tasks). We return to this issue in the last chapter.

1.3.1 The Cell and Its Activities

The cell is the basic unit of life, both structurally and functionally. Its activities include the absorption of nutrients, energy production from the nutrients (and from the sun in photosynthetic cells such as plant cells), interaction with other cells, and reproduction. Many types of cells can perform more specific functions and react to external stimuli.

The activity of cells is cyclical, following a procession called the **cell cycle**. The cycle begins when a new cell is created, progresses through living and growing, and concludes when the cell divides producing two new daughter cells. We will focus on the structure and functions of eukaryotic cells. Eukaryotic cells divide using a process called **mitosis** or using another process called **meiosis**, which produces the gametes that participate in reproduction.

As previously stated, eukaryotic cells contain a **nucleus**, which contains the cell's genetic material. The **cytoplasm,** which surrounds the nucleus, is where most of the living processes of the cell not related directly to processing of the genetic material occur (see Figure 1.1).

The bulk of genetic material is stored in molecules called **DNA**. The data encoded in the DNA is used by the cell to build **proteins**. Proteins are built up from sequences of smaller molecules called **amino acids**. To build a protein, the cell first builds a chain of amino acids. The DNA codes both the identity and the order of the amino acids in the protein. **Enzymes** are an important group of proteins and are responsible for executing the cell's

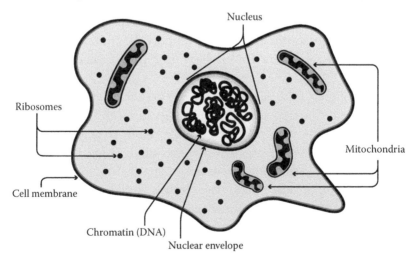

FIGURE 1.1 The general structure of a eukaryotic cell.

activities by participating in chemical reactions. They enable the cell's activities by acting as **catalysts**, that is, by increasing the rate at which chemical reactions occur in the cell by reducing the amount of energy needed for the reaction. A fundamental property of catalysts is that their amount does not decrease due to the chemical reaction they enable, and therefore they can continue participating in further chemical reactions. It is often the case that without a catalyst the rate of reaction is so low that the reaction is virtually nonexistent. As enzymes are organic catalysts, we say that they "perform" the cell's activities.

Chapter 5, which deals with molecular computation, shows how we can make use of biological molecules to implement computational processes. The computations involve, for example, the cutting and splicing of DNA molecules performed by various enzymes. As we will see, some enzymes perform general functions, while others are very specific in their activities and exist only in certain organisms.

The cytoplasm contains **organelles**, which have a variety of functions. One of the most important organelles is the **mitochondrion** (plural **mitochondria**), which is the cell's "power plant." The mitochondrion is responsible for producing energy from sugars by using oxygen. It builds special molecules that are used as a source of **chemical energy** by the cell. In plant and algae cells there exists another important organelle called the **chloroplast**, which is responsible for **photosynthesis**. This is the process whereby CO_2 and water combine into sugar molecules containing energy (e.g., glucose) by using solar energy. This process releases oxygen. Most of the animal kingdom ultimately depends on this process as a source of energy (eating sugars provides energy). Some bacteria also photosynthesize; however, they do not have chloroplasts, and their photosynthesis occurs directly in the cytoplasm.

Another essential role of plants in the living world is that of **nitrogen fixation**. Nitrogen is a building block of amino acids and is therefore necessary for building proteins. Atmospheric nitrogen is relatively inert and does not interact readily with other elements. There are bacteria that fix nitrogen, mainly by a symbiotic process with plants in which soil bacteria fix nitrogen in the roots of plants. The plants synthesize amino acids, which are a source of nitrogen for organisms that feed on plants.

Ribosomes are another important component of the cell. They are responsible for creating proteins based on the genetic information. We will expand on this fundamental process later.

The cell is enclosed in a **cell membrane**, which is composed of many organic molecules, among them proteins and lipids (fats). The membrane is not just a sac containing the cytoplasm but is rather a complex structure that deals with the passive and active transfer of material to and from the cell and is involved in **cell adhesion** and intercellular communication.

The cell's cytoplasm contains other organelles that are responsible for digesting molecules in the environment, that complete the building of proteins after they are synthesized by ribosomes, that deal with cellular division, and more. In the past the cytoplasm was considered to be a disorganized "soup" in which the organelles and other cellular molecules live, but today we know that it has a complex internal structure that determines the location of the different organelles, the regions where cellular processes occur, and more. The internal skeleton of the cell is involved in cellular motion, material transfer in the cell, organelle motion, and even cell division, when the cell divides the cytoplasm and its contents are split between the two daughter cells.

1.3.2 The Structure of DNA

DNA (deoxyribonucleic acid) is a complex chain-like molecule. Its structure was discovered by **James Watson** and **Francis Crick** in 1953 in a major breakthrough in understanding the heredity processes and life in general. We will focus on the properties that enable DNA to be used as data storage used to code for proteins and on the properties of DNA that enable its replication in cell division.

DNA is composed of a long sequence of **bases** or **nucleotides**. The backbone of the DNA molecule is formed by sugars and phosphates to which the bases are attached. There are four types of bases: *adenine* (A); *cytosine* (C); *guanine* (G); and *thymine* (T). One of the major roles of DNA is to code for proteins. We will see later how the sequence of nucleotides determines the sequence of the amino acids that constitute proteins.

One of the first discoveries about DNA was that in every species tested the amount of adenine was equal to the amount of thymine and that the amount of cytosine was equal to the amount of guanine. This is a consequence of the way DNA is organized, but it took a few more years until its exact structure was determined. The model suggested by Watson and Crick explains the equal amounts of adenine-thymine and cytosine-guanine to be the result of DNA molecules being made up of two side-by-side strands. Each strand is a sequence of nucleotides, and each nucleotide in

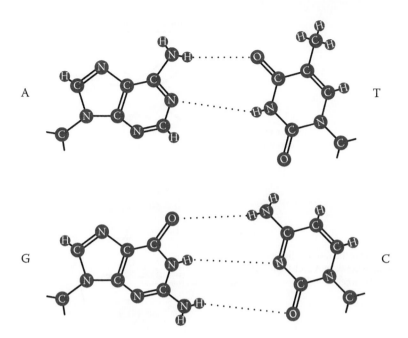

FIGURE 1.2   The structure of the nucleotides and their pairing.

one sequence matches (complements) a nucleotide in the opposite strand. This **base pairing** is such that an A in one strand matches a T in the other strand, and a C matches a G (Figure 1.2). The two strands spiral around each other to form a structure called a **double helix** (Figure 1.3). Each strand has a direction dictated by the orientation of its backbone, and the convention is that each strand runs from the **5'** end (pronounced "five prime end") to the **3'** end (Figure 1.3).

The base pairing is due to chemical bonds (*hydrogen bonds*) between the A and T bases and between the C and G bases. If we suspend two complementary single DNA strands in a solution, they will anneal with each other due to the base pairing and will form a double-stranded DNA molecule.

The exact DNA sequence of each organism is unique. Human DNA contains roughly three billion bases; therefore, the number of possible sequences is beyond imagination. The genetic differences between different species such as humans, chimpanzees, mice, and the tetanus bacteria are characterized by the length of their DNA sequences and the order of their bases. Organisms of the same species also have unique DNA sequences, but the similarity between any two individuals of the same species is very

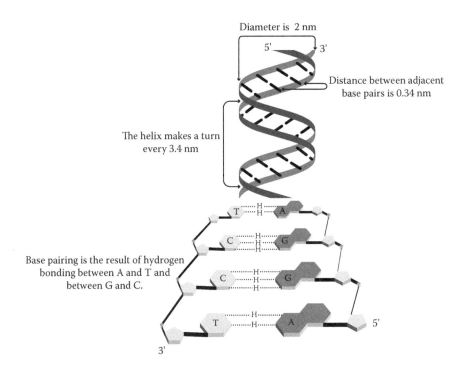

Diameter is 2 nm

Distance between adjacent base pairs is 0.34 nm

The helix makes a turn every 3.4 nm

Base pairing is the result of hydrogen bonding between A and T and between G and C.

FIGURE 1.3 The double helix structure.

high (more than 99% for humans). Nonetheless, these small variations account for the huge variability we encounter.

To summarize:

1. The order of nucleotides in the DNA molecule codes the genetic information stored in the molecule.

2. The two DNA strands match according to the base matching rule: A with T, C with G.

1.3.3 The Genetic Code

After discovering the structure of DNA, the next major challenge was to understand the **genetic code**, that is, to understand how the order of the bases determines the sequence of amino acids in proteins. The research undertaken to decipher the genetic code is a fascinating topic we cannot delve into here, but its main result was that the identity of each amino acid in a protein is determined by a sequence of three nucleotides in the DNA molecule. Therefore, one has to read the DNA sequence as if it was made up of words, each of which contains three nucleotides. Each such word

Second Letter

		T	C	A	G	
T		T T T } Phe T T C T T A } Leu T T G	T C T T C C } Ser T C A T C G	T A T } Tyr T A C T A A } Stop T A G } Stop	T G T } Cys T G C T G A } Stop T G G } Trp	T C A G
C	First Letter	C T T C T C } Leu C T A C T G	C C T C C C } Pro C C A C C G	C A T } His C A C C A A } Gln C A G	C G T C G C } Arg C G A C G G	T C A G
A		A T T A T C } Ile A T A A T G } Met	A C T A C C } Thr A C A A C G	A A T } Asn A A C A A A } Lys A A G	A G T } Ser A G C A G A } Arg A G G	T C A G
G		G T T G T C } Val G T A G T G	G C T G C C } Ala G C A G C G	G A T } Asp G A C G A A } Glu G A G	G G T G G C } Gly G G A G G G	T C A G

Third Letter

Ala = Alanine   Gln = Glutamine   Leu = Leucine   Ser = Serine
Arg = Arginine   Glu = Glutamate   Lys = Lysine   Thr = Threonine
Asn = Asparagine   Gly = Glycine   Met = Methionine   Trp = Tryptophan
Asp = Aspartate   His = Histidine   Phe = Phenylalanine   Tyr = Tyrosine
Cys = Cysteine   Ile = Isoleucine   Pro = Proline   Val = Valine

FIGURE 1.4   The genetic code and the names of the amino acids.

is called a **codon**. With minor exceptions, all living organisms use the same genetic code, and given a codon we can identify the corresponding amino acid. It is easy to see that there are 64 possible codons (4 × 4 × 4); as there are only 20 amino acids, most amino acids correspond to more than one codon. The mapping between codons and amino acids is called the **genetic code** and is shown in Figure 1.4.

The almost complete universality of the genetic code confirms its early origin in the history of life, which is to be expected. On the other hand, one may wonder why the genetic code did not continue to evolve and change in different families of organisms. A possible explanation is that characteristics necessary for survival are intolerant of change, and as a result the original characteristics are preserved and cannot change or improve even if they originated due to "historic accidents." Arguing against this answer is that some organisms have developed a few variations on the genetic code and survived. For example, the mitochondria have their own genome that is different from the genome of the nucleus. The genetic code used by the mammalian mitochondrial genome deviates slightly from the standard genetic code. This leads to the hypothesis that the universality of the genetic code is not only because any change in it will be catastrophic but rather because the genetic code in its current form has evolutionary advantages. The jury is still out on this issue.

Not all of the DNA sequence codes for proteins. A sequence of DNA nucleotides that codes for a specific protein is called a **gene**. We should note that an exact definition of the notion of gene has proven elusive, and the definition of the term has changed as more and more of the complexities of the control mechanisms involved in producing proteins have been discovered. It is worth pointing out that the genes (i.e., the protein coding sequences) are distributed in the genetic material and in higher species constitute only a small fraction of the total genome. There is a lot of active research on the functions of the rest of the genetic material. Sequences that used to be called **junk DNA** and were thought to be useless turn out to have roles in organizing the genome and determining which genes are expressed in the cells. It has been discovered in recent years that many RNA molecules are transcribed from DNA—the first step in protein synthesis—but are not used to build proteins (protein synthesis is discussed in the next section). These RNA molecules (called **noncoding RNAs** or **ncRNAs**) have many roles, and it is already clear from what we know about them that they have a major importance in the regulation of the genome.

1.3.4 Protein Synthesis and Gene Regulation

As previously discussed, the DNA is a double helix containing two strands, and a protein is composed of a sequence of amino acids. When a DNA molecule is to be "read" so that it can be decoded and a protein can be synthesized, the helix is unwound, and the information is read from one of the strands. Here again the base pairing is essential. In the first stage an RNA molecule called **messenger RNA** (abbreviated to **mRNA**) is created. RNA is similar to a single DNA strand—it also contains four types of nucleotides: adenine, cytosine, guanine, and, instead of thymine found in DNA, *uracil* (U), which pairs adenine. We will ignore the chemical differences between mRNA and single-stranded DNA.

To create an mRNA molecule the double helix of DNA is opened at a certain point, and an RNA molecule is built on one of the separated strands using base pairing. This **transcription** process is aided by enzymes called **RNA polymerases** (the process is schematically described in Figure 1.5). It is interesting to note that the transcription can happen on either of the DNA strands, and it is erroneous to think that one strand contains the genetic data and the other strand only complements it. In fact, both

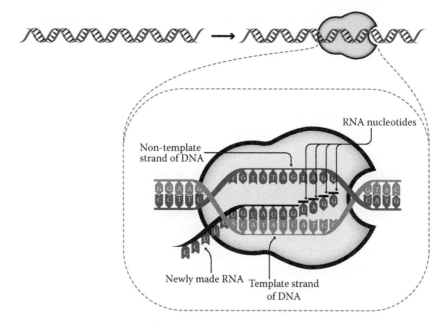

FIGURE 1.5   The transcription process.

strands contain protein building instructions in different locations in the genome. The mRNA transcription process starts at a region called a **promoter**, which marks the beginning of the gene and terminates at a sequence used as the termination signal. The promoter is in front of the transcribed region but is not always directly adjacent to it.

In eukaryotes mRNA transcription happens in the nucleus. After the mRNA is formed, it leaves the nucleus, and protein synthesis can start. The process whereby the information stored in the RNA molecule is converted into a sequence of amino acids is called **translation** and is performed by a cellular complex or machine called a **ribosome**. The ribosome reads the RNA molecule codon by codon (recall that a codon is a sequence of three nucleotides). The mapping of codons to amino acids is represented in the cell by another kind of RNA, known as **transfer RNA (tRNA)**. A tRNA molecule is attached at one end to a specific amino acid and at its other end to an **anticodon**—three bases that are complementary to a codon. When a ribosome is to deal with a particular mRNA codon, the appropriate tRNA molecule attaches to it using base pairing with the anticodon, and the amino acid is detached from the tRNA and attached to the growing sequence of amino acids that will make up the protein. Now the ribosome moves on to the next mRNA codon and so on until the ribosome reaches the **stop codon**. Stop codons do not have corresponding tRNA molecules. The stop codon means that the sequence of amino acids is complete and that the corresponding protein has been synthesized in its entirety. Since protein production is crucial to cells, many thousands of ribosomes in each cell are constantly synthesizing proteins.

Proteins are chain-like linear (nonbranching) sequences of amino acids. Each amino acid (see schematic structure in Figure 1.6(a)) has a part called the backbone, which is virtually identical for all amino acids and a part called side chain (denoted here by R), which is different in each of the 20 amino acids and gives each amino acid its special characteristics. Proteins are the machinery that operates the cell, and therefore most proteins have a unique three-dimensional structure that allows them to perform their function. The linear chain of amino acids folds and twists according to chemical and physical laws after the protein has been synthesized (a schematic structure of a protein can be seen in Figure 1.6(b)). In some cases other proteins are involved in achieving the three-dimensional structure necessary for the protein to function. Some proteins consist of more than one linear chain of amino acids that combine to create the complete protein.

The three-dimensional structure determines the protein's behavior in the cell as proteins interact with each other and with other cellular material according to their physical shape. For example, hemoglobin, which

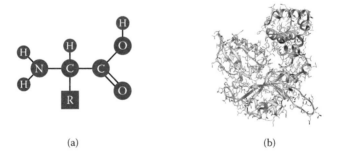

(a)   (b)

FIGURE 1.6   (a) A schematic structure of a single amino acid where R represents the side chain that is different for each of the 20 amino acids. (b) A schematic structure of a protein that is a folded linear chain of amino acids. The ribbon represents the backbone of the amino acid chain with the various amino acids branching from the backbone. (The protein is polymerase β consisting of 355 amino acids.)

carries oxygen in the bloodstream to the muscles, has four cavities that can be loaded with heme, an iron compound that can bind to oxygen.

Researchers distinguish between several levels in the organization of proteins. The first, called the **primary structure,** refers to the sequence of amino acids. The **secondary structure** describes local structures along the amino acid sequence such as different kinds of helices, extended strands that combine to create sheets, turns, and loops. These structures are maintained by a network of chemical bonds (hydrogen bonds) between neighboring amino acids. The three-dimensional structure of a protein is referred to as its **tertiary structure** and describes the three-dimensional location of each atom in the protein.

Determining the tertiary structure of proteins can provide important clues about how they perform their function. Protein structure can be determined using experimental methods such as crystallography using x-rays and nuclear magnetic resonance (NMR). Finding the three-dimensional structure of proteins is an expensive and complex undertaking, and the exact structure of only a relatively small fraction of proteins is known. As the tertiary structure is derived from the amino acids sequence, an ongoing effort exists to build computational tools that will predict the spatial structure for a given sequence. Although in recent years significant progress has been achieved in several aspects of the problem, by and large the current tools have only limited success in predicting the three-dimensional structure of proteins.

To summarize, in the simplest case protein synthesis consists of the following steps. An mRNA molecule is transcribed from the DNA molecule, and the information stored in the mRNA molecule determines the sequence of amino acids in the synthesized protein (translation). The unidirectional information transfer DNA → mRNA → proteins is called the **central dogma** of molecular biology.

As research progressed it became obvious that the central dogma does not cover all possible variations and processes. For example, the genetic material of **retroviruses** (e.g., HIV) is stored in RNA and the virus injects itself into the genome of the host cell. This RNA → DNA process is called **reverse transcription**.

One of the central research topics today is that of understanding the processes that determine which genes will be expressed in the cell (i.e., will be translated into proteins) and at what point in time. **Gene regulation** requires different levels of control. At the most basic level, the DNA molecule contains regulatory regions called **promoters**, which are usually close to the DNA region that codes for a protein. Recall that, for a gene to be expressed, RNA polymerase has to transcribe it to an mRNA molecule. The promoter allows the RNA polymerase to home in on the correct location in the DNA. In eukaryotes, proteins called **transcription factors** are attached to the promoters and are responsible for attaching to the RNA polymerase. Thus, the protein will not be expressed if the promoter is inaccessible (e.g., due to a fold in the DNA). As different cells express different proteins, we find that different promoters are "active" in different cells, and by experimentally attaching a gene to a particular promoter one can cause it to be expressed in certain cells (e.g., in muscle cells or in nerve cells). Cellular regulatory processes are of course much more complicated and are different for bacteria and eukaryotes.

One of the first regulatory structures to be understood was that of the **operon** in bacteria in which a few consecutive genes are jointly regulated (Figure 1.7). In addition to the promoter the operon contains another DNA sequence before the gene or genes, called the **operator**. Proteins, called **repressors**, influence gene expression by attaching to

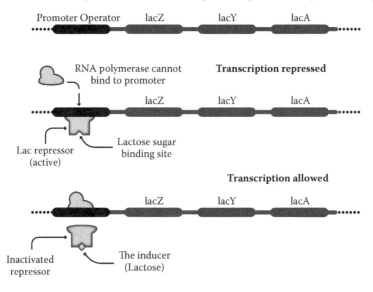

FIGURE 1.7　A schematic structure of the lac operon which controls the metabolism of the sugar lactose depending on the presence of glucose. Three genes lacZ, lacY and lacA are under joint control. The lac repressor is attached to the operator region and thus prevents the RNA polymerase from initiating transcription from the promoter. Once a sugar is bound to the repressor it cannot attach to the DNA and the polymerase can bind to the promoter and transcription is initiated.

the operator and thereby preventing or reducing the transcription rate. Repressors work in two major ways. In one mechanism the repressor is constitutively bound to the DNA and prevents transcription. When an **inducer** molecule is present in the cell it attaches to the repressor molecule, changes its conformation, and prevents it from attaching to the operator and thus enables transcription. In another mechanism, the repressor is not regularly bound to the operon sequence. The repressor protein may exist in the cell but may not be able to attach to the operator unless another molecule is attached to it and changes its conformation. When this happens the repressor can attach to the operator and block transcription. In both mechanisms repressors implement negative control, or inhibition, on gene expression. Other mechanisms implement positive control in which molecules called **activators** strengthen the affinity of the transcription machinery to its target DNA and thus increase the transcription rate of a gene or set of genes. Additional regulatory mechanisms operate on the mRNA and even at the protein level.

1.3.5 Reproduction and Heredity

One of the main characteristics of the living world is reproduction, a process whereby parents create offspring and pass their characteristics on to them. The information transfer from parent to offspring is called **inheritance**. Inheritance exists already at the single-cell level as previously noted: a cell can divide and generate two new cells. An important component of inheritance is the passing on of genetic material in the DNA.

The term **genotype** is used to describe the heritable (or genetic) information, and **phenotype** is used to describe the physical characteristics and behaviors of the organism (which, of course, change over time). Generally speaking, the phenotype is derived from the genotype under the environmental conditions the organism encounters during development. It is common to assume that changes to the phenotype (e.g., breaking a leg, getting the flu) do not change the genotype and are not passed on to subsequent generations (we expand on this issue in Chapter 3). **Genetics** is the biological field dealing with inheritance.

Recall that the genetic information (the genotype) is stored as DNA molecules. In eukaryotes the information is distributed among a few DNA molecules. A DNA molecule together with a few attached protein molecules form a unit called a **chromosome**. The cells of most species contain several chromosomes. The number of chromosomes is the same in all the cells of a multicellular organism, except for the gametes.

Cells go through a number of stages that occur in a fixed order; this constitutes the **cell cycle**. The cycle starts with the formation of a new cell and ends when the cell divides, and each of its descendents starts its own life cycle. The last phase of the life cycle—**mitosis**—is the phase where the cell actually divides. We first focus on the previous phase, in which the DNA is replicated and the chromosomes get duplicated, known as the **replication phase**.

The structure of the DNA molecule as previously discussed lends itself to the possibility of duplicating the genetic information. Just as the synthesis of an mRNA molecule uses one of the DNA strands as a template, a new DNA strand can be created from a previous one by base pairing. To generate a double stranded copy of the DNA molecule, the two original strands separate, a complementary strand is created from each of them, two new double helixes are formed, and cell division can proceed. This process involves many enzymes, of course. Among those are **DNA polymerases**, which synthesize DNA fragments, and **DNA ligases**, which join together the DNA fragments. Errors in replication are one source of **mutations**, random changes in the genetic information.

Not all species reproduce by the mating of male and female individuals. Some species of plants, aphids, and other organisms can reproduce **asexually** in a process whereby a single parent passes a copy of its complete genome to its descendents, similar to the way a single cell divides. Ignoring mutations, these descendents have genetic information that is identical to that of their parent. The situation is more complex in **sexual reproduction**. The genes in the cells of organisms that undergo sexual reproduction are *organized in pairs*, where usually one comes from the father and the other from the mother. We have already noted that different individuals may have different versions of the same gene. A peculiar example is the texture and color of earwax. Two types of earwax in humans are controlled by a single gene (called ATP-binding cassette C11). One version of the gene gives rise to an individual having brown-yellow wet earwax (the form that predominates in Africa and Europe), whereas another leads to gray dry earwax (common among East Asians). The different versions of one gene are called **alleles**. As the two genes in each pair originate in different parents, they may be different alleles. In other words, in sexual reproduction new combinations of alleles are created, giving rise to diverse individuals, and the offspring in a sexually reproducing population will be different from their parents. The differences between the individuals are the raw material for the evolutionary process discussed in Chapter 3.

We now briefly discuss the cellular basis of sexual reproduction. The cells of organisms that reproduce sexually contain pairs of chromosomes, where one chromosome originates in the father and the other in the mother. These cells are called **diploid**, and the number of chromosomes in them is denoted by $2n$. Both chromosomes in a pair are of the same length and have genes for the same characters. They are called **homologous** chromosomes. Human cells contain 23 pairs of chromosomes ($n = 23$) and therefore contain 46 chromosomes.

A noteworthy pair of human chromosomes is the **sex chromosomes**. They come in two forms, the **X chromosome** and the **Y chromosome**, which have different lengths. In females the genotype is XX (i.e., each cell contains two X chromosomes), whereas the male genotype is XY. As males can inherit the Y chromosome only from their father, one can use genetic markers on the Y chromosome to test for paternity and to establish paternal lineage.

An obvious question arising from this discussion is what impact does the difference between alleles in homologous pairs have on a phenotype? The answer is complex and differs from situation to situation. For some properties it suffices to receive the corresponding allele from one of the parents (an allele having this property is considered to be **dominant**), while other properties require getting the appropriate allele from both parents (**recessive** alleles). Returning to the previous illustration, it turns out, for example, that the wet earwax allele is dominant over the dry one.

As a diploid cell contains $2n$ chromosomes, it would seem that the descendant's cells, receiving the chromosomes of both parents, will contain $4n$ chromosomes, and the number of chromosomes will double from one generation to the next. This runaway process does not happen as the genetic information is passed on by specialized cells called **gametes** (in mammals, an egg or a sperm). These are created by a special division process called **meiosis** or **reductional division** and contain only one chromosome from each pair. These cells are called *haploid,* and the number of chromosomes in them is denoted by n. This allows the descendants to have diploid cells, where each homologous chromosome pair contains one chromosome from each parent.

From the current perspective one essential element of meiosis is of particular interest: the mixing of information between the homologous chromosomes. This process is called **crossover** and happens by a pairing up of the homologous chromosomes and exchange of corresponding sections of DNA. The location of the exchange sites is random to a large extent.

If each gene had only one possible molecular form (i.e., a single possible allele), the crossover would have no effect. But as many genes have different alleles, the chromosomes in a homologous pair are not identical; therefore, the crossover process gives rise to new combinations of alleles, producing descendants that have different properties.

After the crossover the cells are still diploid, but their chromosomes are a new combination of alleles. The reduction in the number of chromosomes happens during meiosis when the cell divides into two haploid cells, each of which contains only one copy of the genetic information that has already undergone crossing over. (In reality, this process is more complex. Before the reductional division the genetic information is duplicated. Thus, meiosis can be described schematically as 2n → 4n (duplication) → crossover → 4 haploid cells.) Moreover, note that the haploid cells contain one chromosome chosen randomly from each pair of homologous chromosomes, which have already been recombined using the crossover process. In humans this allows for more than 8 million (2^{23}) possible combinations of 23 chromosomes.

The last stage of sexual reproduction is **fertilization**, when a male gamete and a female gamete combine to create a new diploid cell. So in humans the 23 chromosomes from the father combine with the 23 chromosomes from the mother to create 23 homologous pairs of chromosomes. This process again increases variation as two random gametes, picked from the many gametes available to each parent, participate in the fertilization.

To summarize, sexual reproduction increases variation in the population via several mechanisms:

1. Random crossover.

2. Random selection of one chromosome from each homologous pair.

3. Random selection of the gametes that will undergo fertilization.

In Chapter 3 we will see how it is possible to mimic these various mechanisms to solve **search problems**, in which a solution to a problem is sought from a large space of candidate solutions.

1.4 MODELS AND SIMULATIONS

We mentioned at the beginning of this chapter that using concepts from the biological sciences in computer science can be viewed as "translating" biological phenomena into formal mathematical models. The translations we will deal with in this book are not exact translations but rather the usage of certain elements of the biological phenomena as an inspiration for computational and mathematical ideas.

A **mathematical model** of a system describes the behavior of the system using mathematical tools such as variables, equations, functions, and rules. Historically, it was common to separate models to two types: models described using differential equations that were used to explain continuous deterministic systems and models defined by sets of rules suitable for describing discrete systems that may be nondeterministic. The first type of model might be appropriate for modeling the flow of blood in the circulatory system by a set of equations, whereas models of the second type might be more appropriate for modeling the immune system. In recent years, however, many models are hybrids that combine both techniques.

Many disciplines use mathematical models. They are particularly prevalent in the natural sciences, but mathematical models are also used heavily in social sciences such as economics and sociology. A model focuses on the properties of the system to be studied and describes them formally and exactly. Ideally, this allows for a formal and exact analysis of the system, or at least for gaining a better understanding of the system and its behavior. Models can also be used for predicting the behavior of the system, even when the reasons for the behavior are not well understood. This may help when trying to intervene with a system in order to change its behavior. A case in point is ecological models that can predict the outcome of introducing a new species into an environmental niche and the complex dynamics that might ensue.

After a system is described as a model, it can be handled in two ways:

1. **Finding analytical solutions** of the model: This approach is appropriate when the model is described by equations or functions that are amendable to analytic analysis.

2. **Simulating** the behavior of the system: This is useful when the system is described by a set of rules that are not necessarily deterministic or by equations that cannot be solved analytically. Simulations do not analyze the system mathematically but rather use a mathematical description of the system to simulate its behavior, usually with the aid of computer programs. The simulation calculates the

changes in the variables describing the system according to the rules specified by the model. This is usually done iteratively as a step-by-step process.

When building a model the most important decisions are identifying the variables that characterize the system and its behavior. The model is usually a simplified representation of the system which is tractable; for this purpose, only the properties necessary for describing the particular behavior being studied should be represented in the model.

Here are a couple of examples of models arising in different fields:

- **Consumer behavior**: A simple model of consumer behavior stipulates that the consumer has to choose among n products denoted 1, 2, ..., n whose prices are $p_1, p_2, ..., p_n$ respectively. The consumer is represented by a utility function U, which determines the consumer's *satisfaction* and is a function of the quantities of each of the items the consumer buys. The larger the value of U the more satisfied the consumer is. Note that the function U reflects the particularities of an individual consumer, who might, for example, be an individual who feels the highest satisfaction when owning a small number of necessary worldly goods or a greedy individual who wants the highest quantity of each product he can afford. Each consumer has a budget M, which is used to buy the products. The goal of the consumer is modeled as an *optimization problem,* where U is to be maximized under the constraint that the money spent cannot surpass M.

- **Growth of a bacteria colony**: The next chapter presents a simple mathematical model of the growth of a bacteria colony in the lab. That model does not attempt to be completely precise biologically. The reproduction law we will describe states that the colony grows at a rate proportional to the colony's size at every point in time. If we denote the number of bacteria at time t as $y(t)$, we can derive a simple differential equation that describes the exact behavior of the function $y(t)$. (The solution to such an equation is described in the next chapter.) This function allows us to calculate the size of the bacteria colony without having to resort to simulations. Note that this model is oversimplified and ignores many parameters such as the influence of the amount of available nutrients on the colony's size, the influ-

ence of its density on its growth rate, the minimal amount of time
needed for bacterial reproduction, and more.

- **L-systems**, or **Lindenmayer systems**, are a mathematical formalism based on term rewriting proposed by the biologist **Aristid Lindenmayer** for modeling the growth and development of plants. More recently, L-systems have found several applications in computer graphics. An L-system is specified by a set of rewrite rules of the form $X \to Y$, meaning that every occurrence of the symbol X is replaced with the string Y. The rules of the L-system grammar are applied iteratively starting from an initial state. As many rules as possible are applied simultaneously, per iteration. For example, starting with the string **A** and the rules **A → AB, B → A**, the resulting strings are AB, ABA, ABAAB, ABAABABA and so on (keep in mind that all possible replacements are done simultaneously, so each time the rules are applied the result is a new string that will be the seed of the next iteration). When the strings produced by L-systems are interpreted as graphic commands they can be used to produce striking fractal images, some of them reminiscent of biological patterns.

For example, starting with the string **F** the system composed of the rule **F → F + F − F − F + F**, produces the strings F, F + F − F − F + F, F + F − F − F + F + F + F − F − F + F − F + F − F − F + F − F + F − F − F + F + F + F − F − F + F, and so on. If F is interpreted as the instruction "draw forward" and + and − as a turn of 90° left or right, respectively, the strings can be executed and result in drawings of a variant of the Koch snowflake (see Figure 1.8). By adding more symbols, it is possible to model more complex patterns, for example, patterns of branching growth.

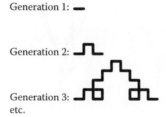

Generation 1:

Generation 2:

Generation 3:
etc.

FIGURE 1.8   The first three generations in the evolution of the Koch snowflake.

The L-system starting with the symbol **X** and the rules **X → F − [[X]
+ X] + F[+FX] − X**, **F → FF** leads to complicated tree structures. The
strings are interpreted as before, with + and − specifying turns of 25°.
The symbols [and] are the magic behind the branching structure.
When executing the string, [is interpreted as a command to store
the current position and angle on a stack, and] is a command to
return to the position and angle of the last push. Finally, note that
the symbol **X** does not affect the drawing. When the string is interpreted as drawing instructions, we can simply ignore the **X**'s. Their
role in the string is to serve as placeholders, allowing the L-system
to keep track of the structure of the curve. Figure 1.9 shows the first
two iterations of this L-system.

These examples of models demonstrate how assumptions about the
behavior of a given system are used to build a model and how they are
formalized during that process. Sometimes these assumptions will allow
only a single outcome, whereas in other cases the hypotheses allow for
a family of common behaviors, all of which adhere to the constraints of
the model. The model allows us to analyze the system and come up with
hypotheses that are used to further investigate the system, either by testing them on the model (e.g., using additional simulations) or by studying
the system directly, back in the lab or in the field. The crucial question of
course is whether the model gives a faithful representation of the system.
This is a complex question for a number of reasons. First, the model simplifies the system and therefore may not allow for a direct comparison of
its variables with the parameters and data that can be measured directly
from the system. Also, when building the model we neglect many of the
system's components, and therefore, even if it behaves in a similar fashion
to that of the system, it is unclear whether the neglected components are

FIGURE 1.9 A simulation of plant growth by an L-system.

crucial for understanding the behavior of the real system. Furthermore, the model may behave accurately in some cases and deviate from the system's behavior in other circumstances, say when the initial conditions are varied. These and other questions all have to be considered when building a formal model that attempts to describe a natural system.

An important point to consider is whether the fact that a model presents a behavior matching the behavior of the system it models provides an *explanation* of that behavior. If our goal is to understand the causes of a particular behavior, a simple simulation will not necessarily suffice. On the other hand, if we aim to find the minimal requirements needed for a system to present a particular behavior, a model may present a satisfactory answer and be considered explanatory. In Chapter 4, dealing with neural networks, we will see that it is easy to build models that present useful and complex behaviors, even though gaining an insight into how the system achieves these goals can be extremely difficult. These models act as black boxes: we define the initial conditions and the rules of behavior and allow the system to self-organize accordingly, thereby often losing the ability to analyze the role of each component in the system, or at least making this task very difficult.

The use of formal mathematical tools to build faithful models of biological phenomena is outside the scope of this book. That field is called **theoretical biology** or **systems biology** and makes use of a wide range of mathematical and computational tools (systems biology is discussed briefly in Chapter 6). It is interesting to note that experimental biologists also take advantage of a completely different kind of models: they perform research on species called **model organisms,** which are easy to use in the lab, in an effort to understand processes also occurring in other species or even across the living world. Often used model organisms are yeasts: particularly the baking yeast (which is a single-cell eukaryote); the microscopic roundworm *C. elegans,* for which the developmental trajectory of each of its 1031 cells has been worked out to amazing detail; the fruit fly *Drosophila,* which has easily observable chromosomes and was used in groundbreaking work in genetics; and mice, which as mammals are closer to humans. Model organisms for plants include *Arabidopsis* and tobacco, which have relatively simple genomes, and maize (corn), whose separate kernels allow for easy observation of the effect of genetic changes on the development of the organism.

This book focuses on attempts to use biological knowledge to develop new methods for solving problems using computers. Clearly, we will need to choose some minimal characteristics of the biological systems that can be used for this purpose. For example, we will see in Chapter 4 how to build a computer system that is capable of independent **learning**, thereby eliminating the need to reprogram it to solve new problems. In Chapter 3 we will harness some basic insights into the evolutionary process to create computer programs that solve difficult **optimization** problems (i.e., problems of finding a maximal value of a multivariable function that is not differentiable or does not have a simple formal description) by generating a collection of possible solutions and applying evolutionary like processes to promote the best ones. The main consideration in building such models is their usefulness as computational tools and not necessarily their precision as descriptions of biological systems or processes. This is the reason this field is often referred to as **biologically inspired computing**.

Building and using such models allows us to develop an intuition about biological processes, even though the biological systems are much more complex, contain many interacting mechanisms, and behave in diverse ways in different species under different conditions. This is apparent even when looking at the previously described biological systems. The general understanding of the biological systems that is the starting point of the techniques we describe does not necessarily match the real behavior of the biological processes, nor is our goal to achieve a better understanding of the biological behavior. Having said that, using the models may help us in understanding general biological principles rather than specific processes. In the next chapter we will see how mathematicians and computer scientists attempted to construct models that explain the amazing ability of living things to self-replicate, that is, to generate offspring similar or identical to themselves. These researchers were trying to understand which requirements are necessary to build a self-replicating system and were not attempting to understand how self-replication occurs in nature. Surprisingly, after the discovery of DNA and its behavior it turned out that the abstract model studied by theoreticians was similar in fundamental respects to what happens in living cells.

We have stated already that other computational tools, not inspired by biology, make use of the basic principles discussed in this book. These principles include self-organization, local interactions between components, asynchronicity, redundancy, use of noise, and nondeterministic and parallel and distributed systems. The reader who studies the models we describe in this book and the ways to analyze, implement, and test them (e.g., varying parameters and determining their influence, attempts to combine different models, analyzing the models' fault tolerance) will gain important insights and intuition about the relative merits of these principles that are also used in the development of large software systems not necessarily based on biological models.

In Chapter 5 we will discuss the usage of biological molecules (DNA and proteins) to perform computations. This obviously does not involve modeling a biological system but can be rather seen as the opposite—using biological molecules to model the computational process.

1.5 SUMMARY

The chapters ahead discuss several paradigmatic computational models and techniques that were inspired by observing the living world or by biological knowledge. In each chapter we try to give the relevant biological background and intuition and to present the fundamental aspects of the computational approaches we discuss. We do not assume any prior biological knowledge, and each chapter is self-contained.

In Chapters 2–5 we describe topics that by now are considered well established and form the core of the emerging field of biological computation. In the concluding chapter, Chapter 6, we provide a survey of additional topics like swarm intelligence, artificial immune systems, artificial life, formal languages to describe biological systems, and system biology. These approaches interact with the approaches described in the previous chapters and expand on them and show which directions this field may develop in the future.

The techniques discussed in Chapters 2 through 4 and in Chapter 6 can be applied immediately to solving practical problems, whereas at the moment the notion of molecular computing discussed in Chapter 5 is of less immediate use to programmers. All the models discussed, however, provide insight about biology, about the nature of computation, and about how the two fields relate to each other.

Each of the techniques we discuss exists in many varieties and can be extended in various ways for different purposes. Each chapter includes references to further reading, and the final chapter includes a list of recommended books. We hope that, after studying the material presented in this book, the interested reader will be well prepared to independently explore the wealth of useful and related material available online.

Our goal in each chapter is to present the fundamental concepts behind the techniques and examples of types of applications for which it may be relevant. We hint at some of the directions in which each technique may be modified, often via the exercises that appear at the end of each chapter. We intentionally refrain from presenting most of the material formally in order for the text to be accessible and readable. Several fundamental theorems, however, are presented formally along with a proof or a sketch explaining how the proof of the theorem is constructed.

The techniques presented in the following chapters are powerful and easy to implement, and many free implementations (in both senses of the term) exist online. We encourage readers to "play along" and experiment.

Enjoy the ride!

1.6 FURTHER READING

The following introductory textbooks in biology can be used to supplement the short introduction to biology we present in this chapter.

Campbell, Neil and Jane B. Reece. 2009. *Biology*, 8th ed. San Francisco: Pearson Benjamin Cummings.

Solomon, Eldra, Linda Berg, and Diana W. Martin. 2007. *Biology*, 8th ed. Florence, KY: Thomson Brooks/Cole.

Starr, Cecie, Ralph Taggart, Christine Evers, and Lisa Starr. 2008. *Biology: The Unity and Diversity of Life*, 12th ed. Florence, KY: Thomson Brooks/Cole.

1.7 EXERCISES

Exercises with solutions are marked with bold numbers.

1.7.1 Biological Computation

1. When building a system one tries to avoid the existence of **single points of failure**, a failure which will cause the system to crash. Use this notion to discuss why the control mechanisms of the non-standard computational models (as summarized in Table 1.1) may account for the systems' robustness.

2. Try to use your answer to the previous question to discuss how self-organization can contribute to robustness. What are the dangers in relying on such a mechanism?

3. Discuss which of the properties of nonstandard computation enumerated in Table 1.1 will make standard programming techniques more difficult. Consider all the stages of building software systems: analysis, design, programming, testing, and maintenance.

1.7.2 History

4. Which of the following constitute positive feedback and which constitute negative feedback?

    a. Feeling cold, the body reacts by shivering designed to raise its temperature.

    b. Global warming causes the glaciers to melt. As glaciers reflect a lot of sunlight, their existence reduces warming.

    c. Fat cells secrete the hormone **leptin**. When the amount of fat decreases, less leptin is secreted, and that causes brain cells to send hunger signals. Cells in the gastrointestinal tract can determine how much food has been consumed and can signal other brain cells to stop the eating. The newly replenished fat cells resume the leptin secretion, and the feeling of hunger passes.

1.7.3 Biological Introduction

5. How many bits are needed to code the information contained in a single nucleotide?

6. Does the double-stranded DNA molecule contain more information than each of the separate strands?

7. Some regions of DNA contain repeating long sequences of the nucleotides AT (i.e., regions of the form ATATAT). Explain why these sequences are believed to have a role different from that of storing genetic information.

8. Try to come up with biological reasons that could explain why the genetic code is not a one-to-one mapping (i.e., why there are multiple codons that code the same amino acid). Does your hypothesis lead to any predictions that can be tested either by examining the genetic code or experimentally?

9. Where is the genetic code stored? Try to figure out how the genetic code becomes expressed in the cell.

10. We have noted in Section 1.3 that it is believed that a change in a phenotype does not imply a corresponding change in the genotype. How is this assumption manifested in the formula "DNA → RNA → protein" (the **central dogma** of molecular biology)?

11. In the following table, match the items in column A (biological terms) with items in column B (computational terms). Note that not all terms need necessarily be matched, and more than one term can be matched to some of the items.

Column A	Column B
DNA	Hardware
Protein	Software
Enzyme	Programming language
Genetic code	Compiler
Ribosome	Machine language

How successful is the analogy between the two domains?

12. Assume eye color has two possible alleles: A for brown eyes; and a for blue eyes. An individual with the alleles A and a (denoted by Aa) will have brown eyes (i.e., a does not affect the eye color in this case). We say that in this situation A is **dominant** relative to a, and a is **recessive** relative to A. Assume two parents are both Aa. Every child will get one gene from each parent (A or a) with probability of one half. Calculate the probability of each possible genotype for the descendents and the probability of each phenotype (i.e., eye color).

13. Assume the gene for leaf color has two possible alleles: A for red leaves; and a for white leaves. An individual with both the A and a alleles (denoted as Aa) has pink leaves (which is the combination of red and white). Assume two parents are both Aa. Every child will get one gene from each parent (A or a) with probability one half. Calculate the probability of each possible genotype for the descendents and the probability of each phenotype (i.e., leaf color).

14. We discussed two processes that use DNA as a template for creating a new molecule: creating mRNA and creating new DNA. In both processes, errors can occur, as is expected in any chemical process. The cell deals with such errors by having error correction mechanisms.

   a. In which of the two processes are errors more critical? What hypothesis can you deduce from this regarding the error correction mechanisms?

b. Most proteins are generated in the cell in many copies and thus are transcribed again and again. Does this change your previous answer?

15. We have discussed the fact that most cells in multicellular organisms, except for the gametes and a few types of specialized cells, contain the same genetic information. Cells go through a specialization process and have different roles. Discuss how *regulatory* mechanisms could be employed toward this end and how regulation has to integrate with cell division.

1.7.4 Models and Simulations

16. You may have encountered *queueing theory* in the past (e.g., in a computer networks course). Which of the types of models we discussed is most similar to the models of queueing theory? How does queueing theory allow for analytical solutions, despite the fact that the models are based on random behavior (i.e., they are probabilistic models)?

17. You may have encountered *game theory* in the past. Which of the models we discussed is most similar to the models of game theory?

1.8 ANSWERS TO SELECTED EXERCISES

2. If the programmer foresaw the possibility of a particular fault that can crash the system and determined how to deal with it, the system may be able to recover from it. Alternatively, a self-organizing system may be able to recover from many kinds of faults. The more complex the system and its possible set of problems, the odds of planning for all faults in advance decreases, and therefore self-organizing capabilities may become more important. This gives rise to the danger of the system recovering (self-organizing) in an inappropriate way. The best way of dealing with this is testing how the system reacts to an array of problems and adjusting the system if it reacts inappropriately to problems.

4.

a. Negative feedback: shivering raises the body's temperature, which decreases the feeling of cold, which causes the shivering to stop and the system to revert to its initial state.

b. Positive feedback: the glacier melting reduces the solar reflection, and therefore Earth's temperature rises. This causes more glaciers to melt, and the warming trend increases.

7. Easily compressible data cannot contain much information (as the same data can be represented in a more concise fashion). The more compressible the data, the less information it carries. A long repeating sequence can be easily compressed ("repeat AT 200,000 times"). Therefore, it would seem that repeat sequences contain very little information. If such sequences exist and do not disappear during the course of evolution (see Chapter 3), it might suggest that they have a role other than that of storing information (e.g., a structural role).

9. The genetic code is stored by the set of tRNA molecules in the cell—molecules whose one end presents an anticodon and whose other end is linked to the appropriate amino acid. The tRNA molecules are coded for in the DNA of the cell. Loading the tRNA molecules with the appropriate amino acid for the anticodon is the function of specific enzymes called *tRNA synthetases*, which themselves are coded for by DNA genes, of course. Try to figure out how the genes involved in this process can themselves become expressed: how does the process get started?

CHAPTER 2

Cellular Automata

C ELLULAR AUTOMATA (CA) WERE proposed in the 1950s by the famous mathematician **John von Neumann** as a model for studying the ability of organisms to self-replicate. Since then, the CA model has been used to describe many phenomena in diverse research areas. Some of these areas are biological and include models for the spreading of diseases and the behavior of bacteria colonies (see Section 2.1), but CA are also used in nonbiological fields, for instance for creating physical simulations. CA can be enjoyed as purely recreational mathematics, but we will use this model to discuss deep topics in biological computation.

2.1 BIOLOGICAL BACKGROUND

2.1.1 Bacteria Basics

The vast majority of living organisms are the **prokaryotes** of which there are two types, **Bacteria** and **Archaea**. The prokaryotes are characterized by the absence of a nuclear membrane and consequently the noncompartmentalized nature of their single cell. All organisms synthesize proteins on molecular complexes called **ribosomes**, and therefore the need to accommodate a large number of these relatively large molecular machines means that the minimum size of autonomous organisms is generally a few hundred nanometers (1 nm = 10^{-9} of a meter) across; however, they can be as small as 200 nm in diameter. Figure 2.1(a) shows an electron microscope image of a colony of the *Vibrio cholerae* bacteria, which infect the digestive system; a schematic description of bacteria's structure is given in Figure 2.1(b).

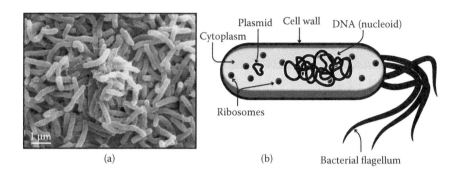

FIGURE 2.1   (a) A colony of *Vibrio cholerae* bacteria; (b) A schematic structure of a bacterium.

Bacteria exhibit an amazing range of sizes, metabolic capacities, and lifestyles, with the largest characterized representative exceeding 500 micrometers ($1\ \mu m = 10^{-6}$ of a meter) in diameter. This remarkable bacterial diversity not only is fascinating for bacteriologists but also is crucial to the continued existence of our world. For example, various species of bacteria release oxygen into the atmosphere, whereas others, living inside the human body, influence the delicate balance between health and disease.

The most notorious property of bacteria is their capacity to reproduce rapidly. In a rich medium (containing sugars and amino acids), the oft-studied *Escherichia coli* divides every 20 minutes to produce 72 generations per day. Such growth, if left unchecked, would generate a mass of bacteria equal to the mass of the earth in two days. In general, bacterial growth involves replication of genetic material followed by binary fission into two identical daughter cells, each containing one copy of the genome.

2.1.2 Genetic Inheritance—Downward and Sideways

In contrast to the linear chromosomes of eukaryotes, the essential component of a bacterial genome is typically a single, closed circle of double-stranded DNA, 4–5 megabases (Mb) long, called the **bacterial chromosome**, which is compacted inside the cell to form a structure called a **nucleoid**. In addition, the genomes of some bacteria also contain smaller circles of double-stranded DNA, known as **plasmids**, which range in size from 1000 bases to several megabases. Essential bacterial genes, those required for growth and reproduction, are generally encoded within the chromosome, whereas genes needed only under exceptional conditions are encoded within a plasmid. For example, genes encoding resistance to antibiotics are often encoded in plasmids. A very large fraction of the bacterial chromosome DNA (about 85% in *Escherichia coli*) encodes for proteins, whereas for humans less than 5% of DNA encodes proteins (see Chapter 1).

The replication and equal partitioning of the bacterial genome that occurs during bacterial growth underlies the classical vertical inheritance of traits from one generation to the next. However, bacteria can acquire genes in another manner, termed **horizontal gene transfer** or **lateral gene transfer** (**LGT**), whereby genetic material is received by one bacterium from an unrelated bacterium. Comparative genomic analysis indicates that LGT impacts bacterial evolution in the long-term and additionally serves as a pathway for acquiring transitory traits, such as the reduced susceptibility to antibiotics, a worrying and increasingly prevalent phenomenon since the 1960s.

2.1.3 Diversity and the Species Question

Classifying bacteria into species and strains has been a challenge since they were first visualized and is even more so presently, despite the avalanche of genomic information becoming available each day. This problem is particularly pertinent as millions of bacterial species are thought to exist in our environment that have not yet been sampled or grown successfully in a laboratory.

It is becoming clear that bacterial genomes are composed of **core sequences**, such as essential genes; **dispensable sequences** characterized by their erratic appearance and nucleotide variability across a **panel of isolates** (the same bacterial species sampled repeatedly and independently); and **strain-specific genes** that are unique to a given isolate. These observations have created the notion that a bacterial species cannot be represented by a unique genomic sequence but instead is defined by a **pan-genome**, which is the core sequences plus the collection of dispensable and strain-specific sequences. Moreover, since bacteria exist in nature in a particular environmental niche, such as inside our guts or in the ocean, an initially radical research approach is to study the niche community as an entity and to call the genetic repertoire of the diverse microbes therein a **metagenome**. In recent years, metagenomics has become a mainstream field. Within such an environment, events of LGT are expected to occur quite often. Currently, more than 100 metagenomic projects are under way for which novel bioinformatics tools are being developed. Together with the insights gained from about 700 fully sequenced microbial genomes, our understanding of what really defines a bacterial species is likely to improve.

2.1.4 Bacteria and Humans

Currently, researchers are analyzing the metagenome of the human gut, estimated to contain some 10 trillion individual bacteria that are members of more than 1000 different species. Bacteria are found also in other human habitats, including the female reproductive tract, the skin, and the mouth. Together, all the bacteria in the human body constitute the human **microbiome**. Bacteria living inside the body that do not cause disease are known as **commensals**, whereas those associated with human disease are called **pathogens**. Bacteria play crucial roles in the life of higher organisms like humans. Many bacteria develop symbiotic relationships with their hosts. For example, bacteria participate in key stages of food digestion in humans. Furthermore, the mere existence of the "good" bacteria keeps at bay the number of the pathogenic bacteria because they compete for the same environmental resources. However, the distinction between harmless or even beneficial bacteria and pathogens is becoming blurred as emerging studies reveal that many recognized pathogens, such as the bacteria responsible for pneumococcal diseases, are commonly carried asymptomatically. Such studies highlight the gaps in our present understanding of pathogenicity and of the interplay between different species and strains of bacteria that cohabit at the same niches.

A better understanding of pathogenicity is urgently needed to expand our ability to combat bacterial diseases. Antibiotics have changed the outcome of the everlasting war between humans and bacteria dramatically in the last 60 years. However, the excessive use of antibiotics brought about the emergence of "superbugs," such as the multidrug resistant *Klebsiella pneumoniae* and methicillin-resistant *Staphylococcus aureus* (MRSA), which do not respond to antibiotics and pose a major public health risk.

2.1.5 The Sociobiology of Bacteria

A growing body of studies demonstrates that bacteria produce and secrete signaling molecules that other bacteria detect and to which they respond, for example, by changing gene expression. These signals enable bacteria to exhibit advantageous communal behavior, which appears to underlie several key phenomena, such as **biofilm** formation, whereby bacteria organize into a particular architecture. Another phenomenon is **quorum sensing**, a mechanism by which bacteria can estimate the density of their colonies by monitoring the amount of signal molecules secreted by the community members. In fact, a colony of bacteria can be regarded almost as a multicellular organism rather than a collection of individual cells.

An impressive example of such collective behavior is seen in Figure 2.2: an expanding colony of bacteria encounters a fragment of glass wool on the surface of the petri dish, causing the individual cells in the colony to

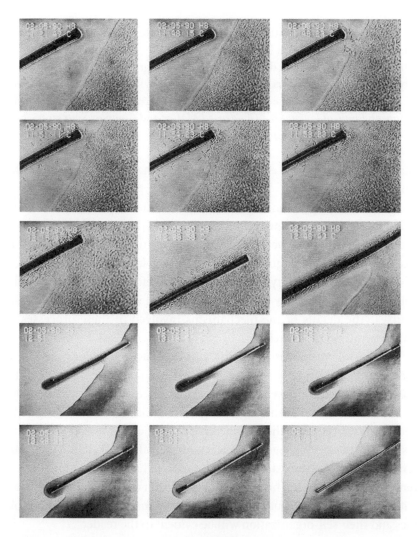

FIGURE 2.2   A series of pictures showing how a fiber approached by a bacteria colony is being surrounded by an individual cell that leaves the colony and engulfs the fiber. Eventually, the coated fiber is being absorbed into the colony. (From Shapiro and Dworkin, Eds., *Bacteria as Multicellular Organisms*, Oxford University Press, 1997. With permission.)

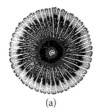 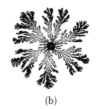 

(a)　　　　　　　　　(b)　　　　　　　　　(c)

FIGURE 2.3   Branching patterns exhibited by *Paenibacillus dendritiformis* bacteria: (a) shows the pattern at higher food levels; (b) shows the typical pattern with intermediate levels of food depletion; (c) shows the growth for a very low level of food. These forms of bacterial self-organization provide the colony with the ability to make an efficient use of the available resources. (From Ben-Jacob, Eshel, *European Physical Journal B – Condensed Matter and Complex Systems* 65, no. 3, 315–322. With permission.)

change behavior in such a way that the *colony* manages to engulf the glass wool fragment.

Another example concerns pattern formation by bacteria colonies. **Eshel Ben-Jacob** has shown (Ben-Jacob, 2007) that a colony of the same bacteria can grow in a very different patterns under different conditions like food availability (Figure 2.3). Furthermore, it was shown that a colony of bacteria has the ability to "learn," and therefore its response to a first course of antibiotics may be very different from its response to successive courses. The details of the mechanisms involved in these processes are only beginning to unravel.

The "social" processes occurring within a bacterial colony are so rich that ongoing studies are investigating whether certain forms of cooperative multicellular behavior even lead to the emergence of "cheaters," individual bacteria that reap the benefits of cooperation without contributing to the community. In general, the recent coining of various phrases by bacteriologists (e.g., the sociobiology of bacteria, microbial multicelluarity, quorum sensing) illustrates the extent to which this area has become a focus of current bacterial research. The following description of cellular automata deals with artificial systems, but we are sure that the analogy with bacterial colonies and their self-organization will be evident to the reader.

2.2 THE "GAME OF LIFE"

We will start our discussion of cellular automata by describing an example of a two-dimensional CA known as the "Game of Life." A more general and formal definition of CA will be presented later.

A two-dimensional cellular automaton is a square grid of cells, each of which is in one of a finite number of states. The automaton progresses from one generation to the next using the following procedure: each cell inspects its state and the states of its neighbors and updates its state using a simple rule. The same rule is used by all the cells. A configuration is a collective state of the automaton, that is, a description of the state of each cell in a given generation.

The best-known cellular automaton is the "Game of Life" (or Life for short) presented by John Horton Conway in 1970. In Life, the grid is infinite in both dimensions. Each cell can be in one of two states: alive or dead (equivalently, full or empty). The update rule depends on the number of live cells in the immediate neighborhood of each cell (i.e., how many of the eight cells around a given cell, including diagonals, are alive):

Birth rule: a dead cell with exactly three live neighbors comes to life; in any other case a dead cell remains dead.

Survival rule: a live cell with two or three live neighbors stays alive.

Death rule: a live cell with four or more live neighbors dies of overcrowding; a live cell with at most one live neighbor dies of loneliness.

Note that the states of the cells are inspected before the update. That means that first all the cells needing to be updated are identified, and then the updates happen simultaneously for all of the automaton's cells. The rules are demonstrated in Figure 2.4.

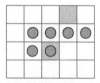

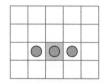

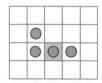

The lower gray cell's neighborhood consists of 4 live cells so it will die in the next generation.
The upper gray cell's neighborhood consists of 3 live cells so it will change to live in the next generation.

The gray cell's neighborhood consists of 2 live cells so it will stay alive in the next generation.

The gray cell's neighborhood consists of 3 live cells so it will stay alive in the next generation.

FIGURE 2.4 Examples of the execution of the transition rule of the "Game of Life" (a live cell is marked with a filled circle).

Typically, we will use a computerized simulation to follow the evolution of cells in the "Game of Life," but we could also run the simulation manually:

1. Start with an initial configuration where live cells are represented on a checkerboard using the black pieces.

2. Find all the cells that will die in the next generation. Mark each such cell by adding another black piece on top of the one in the cell.

3. Find all the cells that will be born in the next generation. Mark each such cell by placing a white piece on it.

4. Make sure that you did not miss anything; then update the board by removing all pieces from the dead cells (those with two black pieces), and execute the birth rule by replacing all the white pieces with black pieces.

5. Now that you have the next generation; repeat from step 1.

Figure 2.5 shows a few configurations that lead to interesting patterns you may want to follow for several generations.

Studying these examples demonstrates that although the rules are simple, the system's behavior can be complex and hard to predict. Some initial configurations never change when the rules are applied; other initial configurations cycle between a finite number of configurations (i.e., these configurations are called **oscillators**); still others preserve their shape but move over the board.

Observing Life can have a hypnotic effect, but it also raises many questions such as the following:

- Can we predict the board's configuration in n generations without executing the game for n generations?

The clock

The glider

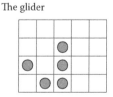

FIGURE 2.5   Initial configurations that lead to interesting behavior.

- Do all initial configurations lead to a steady state, after which there are no further changes?

- Do all initial configurations lead to a state with a fixed number of live cells, that is, a state after which the automaton will not grow further?

Conway's goal was to find simple rules for cellular automata with the following elusive properties:

- There will be no initial configuration for which it will be easy to prove that the population of live cells will grow indefinitely without bounds.

- There will be initial configurations for which it will *seem* as if the population of live cells grows without bounds.

- Some initial configurations will change during many generations, but eventually they will settle into one of three situations: (1) extinction (due to overcrowding or loneliness); (2) periodic oscillation between several configurations; or (3) achieving a steady state.

In this regard, one of the most interesting examples is the "R-Pentomino," which is an innocent-looking pattern of five cells shown in Figure 2.6. It takes 1103 generations for it to stabilize. Readers are encouraged to try it on their favorite Life applet (e.g., at http://www.ibiblio.org/lifepatterns/) and be amazed by the richness of the final pattern. No wonder this pattern is known as one of the *Methuselah* patterns, named after the biblical Methuselah who lived for 969 years.

From this, and many other such examples, it became clear that indeed, as Conway intended, predicting the behavior of the system is extremely complicated. In fact, it turns out that it is *impossible* to predict the evolution of the system from its initial configuration, and its evolution can be determined only by simulating the game. This can be proved mathematically

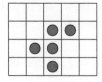

FIGURE 2.6   The R-Pentomino. This simple pattern takes 1103 generations to settle down into a complicated "community" of blocks, blinkers, gliders and more.

by showing that Life can function as a computer (in formal terms, it is equivalent to a universal *Turing machine* [TM]). Therefore, predicting the fate of an initial configuration in the "Game of Life" is equivalent to deciding whether a computer program will halt for a given input or what its output will be, which are known to be undecidable problems. We will return to this important point later in the chapter.

2.3 GENERAL DEFINITION OF CELLULAR AUTOMATA

So far, we have discussed a very specialized case of a cellular automaton. We will attempt now to generalize the properties of a CA.

The cells may be in one of a larger set of states (not just "live" or "dead" as in Life). The set of states has to be finite (typically the set is rather small and contains no more than 10 states). The set of states is usually denoted by the letter Σ and the number of states by k.

The cells consulted for each cell by the behavior rules are called the **cell's neighborhood**. The neighborhood may be different from one automaton to the other. Sometimes the neighborhood is defined to be the cells perpendicular to the cell (four cells in the two-dimensional case, called the **von Neumann neighborhood**). In other cases the neighborhood contains all the adjacent cells (eight in the two-dimensional case, called the **Moore neighborhood**). In general the neighborhood may include cells that are not directly adjacent neighbors and may have any shape.

An automaton need not be two-dimensional; it may be one-dimensional (a linear formation of cells) or three-dimensional (a cube of cells) or have higher dimensionality. The cells need not be on a square, chess-like grid. Any regular tiling of the automaton space such as triangular or beehive shaped tiling will do. We will assume that the automaton space is infinite. When implemented by a computer program, we will usually enlarge the board when necessary. But we may also assume other boundary conditions, such as a two-dimensional surface on a sphere where the edges meet or a board laid out on a torus.

Taking all of this into account, we see that the elements needed to define a particular CA are as follows:

- The **layout** of the board (e.g., a two-dimensional board, a beehive) and its boundary conditions.

- The **set of states** Σ. At any instance every cell is in one of these states. We will denote the state of cell i at time t by S_i^t. A configuration is a collective state of the automaton's cells, that is, a description of the state of each cell in a given generation.

- The **neighborhood** of every cell. The states of all cells in a cell's neighborhood are used as input when calculating the state of the cell in the next generation. The neighborhood is denoted by N_i^t, which specifies the states of the cells in the neighborhood of cell i at time t. When we are not interested in a specific time period the t may be omitted.

- The **transition rule** determines the state of each cell in the next generation, based on its current state and the current states of the cells in its neighborhood. For simplicity we will assume that a cell's neighborhood contains the cell itself. We will denote the transition rule by $\delta(N_i^t)$. Note that the transition rule may not consider the location of a specific cell on the board (e.g., its x and y coordinates). Using such information would violate the principle that each cell can get information only from its neighbors and cannot make use of "system-wide" properties (i.e., a cell can use only local rather than global information). This is the basis of the cellular automata model.

So, a particular automaton is defined by four properties: (1) the board; (2) the set of states; (3) the definition of a cell's neighborhood; and (4) the transition rule.

All the cells' neighborhoods have the same shape and this shape does not change with time. For now we will also assume that all the cells transition at the same time to their next state. This is called a **synchronous transition**. Accordingly, the time in the system can be divided into distinct and discrete time units, a regime called **discrete time**.

Note that discrete time is very different from the notion of time as continuous, and the incompatibilities between these two notions of time can lead to paradoxical behavior as in Zeno's arrow paradox. Imagine an arrow in flight. Now suppose time is divided into a series of indivisible moments. At any given moment, the arrow is at an exact location, so it is not moving. But if at any moment in time there is no motion, we must conclude that movement cannot happen—that the arrow is motionless. The paradox actually stems from the idea that time can be divided into discrete units. Thus, we must be careful when we model dynamic systems using CA not to get into situations when the discrete nature of time might be a confusing factor in the behavior of the simulation or to assume that a discrete time model necessarily captures all the important elements of a continuous phenomenon (or vice versa, of course).

A state s for which $\delta(s,\ldots,s) = s$ is called a **quiescent state**: a neighborhood where all the cells (including the cell itself) are in a quiescent state and will remain in a quiescent state at the next time step. The quiescent state of cells in the "Game of Life" is "dead": the rules of the game will not result in a creation of any living cell if the entire board is empty.

We summarize this section with a formal definition of a CA as follows:

FORMAL DEFINITION OF A CELLULAR AUTOMATON

A d-dimensional cellular automaton **A** is a 4-tuple (Z^d, Σ, N, δ) where:

- Z^d is the description of the space defining the automaton's layout.
- Σ is a finite set of the possible cell states.
- N is an ordered subset of Z^d of size $n+1$ called the neighborhood of **A**. For a cell $x \in Z^d$, the neighborhood of x is defined to be the cells in positions $x + r_i$ for $i = 0,1,\ldots,n$. where r_i is a vector in the d-dimensional space ($i = 0$ represents the cell x itself).
- δ: $\Sigma^{n+1} \rightarrow \Sigma$ is a function called the transition rule of **A**.

This is a standard definition of a cellular automaton. Some variations later in this chapter do not adhere to this definition.

2.4 1-DIMENSIONAL AUTOMATA

A one-dimensional (1-D) automaton is a tape or sequence of cells. For example:

is an instance of a 1-D automaton with the state set $\{0,1\}$ where the tape is of length 4. We often assume that the tape is circular (i.e., the right-hand neighbor of the rightmost cell is the left cell and vice versa). Typically, a cell's neighborhood will be described by the radius r around the cell. Therefore, the number of cells in the neighborhood is $2r + 1$ (r cells on each side of the cell and the cell itself).

If we keep the number of states and the size of the neighborhood moderately small, the number of different possible neighborhoods also remains relatively small, and we will be able to describe the automaton using a simple table that holds the next state of a cell as a function of the states in its neighborhood. For example, an automaton where $k = 2$ and $r = 1$ is an automaton in which each cell can be in one of two states and the next state is determined by the states of the cell and its two neighbors.

Table 2.1 describes a possible transition rule for such an automaton. Note that the entire table represents a single transition rule. Also note that this table enumerates all possible states of the environment.

If the automaton's initial configuration is as previously described

1	0	0	1

After applying the transition rule, its state in the next generation will be

1	0	1	1

Observe that all tables describing transition rules for 1-D automata with $k = 2$ and $r = 1$ have eight rows. Note that the number of rows in the table is determined by the number of possible neighborhoods (here $2^3 = 8$) and is not dependent on the size of the automaton, namely on how many cells it contains. For simplicity, the following notation has been devised: the next state of cells as a function of the neighborhoods will be written as a sequence of bits (read bottom to top from the "Next State" column in Table 2.1). So the transition rule in Table 2.1 can be described by the sequence 01101110. This sequence is the binary representation of the decimal number 110, so this transition rule is known as *Rule 110* (Table 2.2). This shorthand for describing the transition rules is known as **Wolfram numbers**, after Stephen Wolfram (2002), who first proposed and used this notation.

TABLE 2.1　Rule 110

	Neighborhood		
State of Left Neighbor	Current State of Cell	State of Right Neighbor	Next State
0	0	0	0
0	0	1	1
0	1	0	1
0	1	1	1
1	0	0	0
1	0	1	1
1	1	0	1
1	1	1	0

TABLE 2.2   Rules and Their Decimal Names

Converting a Rule to Its Decimal Name								
Neighborhood as a binary number	111	110	101	100	011	010	001	000
2 to the neighorhood's state	128	64	32	16	8	4	2	1
Next state		0	1	1	0	1	1	1	0

$$01101110_b = 128 \cdot 0 + 64 \cdot 1 + 32 \cdot 1 + 16 \cdot 0 + 8 \cdot 1 + 4 \cdot 1 + 2 \cdot 1 + 1 \cdot 0 = 110_d$$

As every eight-digit binary number can represent an automaton's transition rule, we see that the number of possible automata is $2^8 = 256$. This is a significant observation since it means that we have now a systematic way to address all possible transition rules for binary 1-D automata. Furthermore, practically speaking, we can reduce this number significantly. There is an obvious symmetry between the digits 0 and 1. In addition, we will usually consider only automata with a quiescent state (i.e., for the configuration where all the cells in the neighborhood are 0, the next state of a cell in state 0 remains 0). A further simplification may arise from allowing only rules that are symmetrical around the middle cell in the neighborhood. In this case, we will want the neighborhoods 100 and 001 to have the same value as well as neighborhoods 110 and 011.

The first simplification reduces the number of possible automata by half. The second simplification implies that the binary representation of a transition rule ends with a 0, since this represents the transition $000 \rightarrow 0$. The third simplification allows us to look only at the values of five neighborhoods to infer all the other values of the transition rule. Taken together, we see that the interesting transition rules are of the form $a_1 a_2 a_3 a_4 a_2 a_5 a_4 0$ (note that a_2 and a_4 appear twice due to symmetry), and the number of rules to consider for automata with $k = 2$ and $r = 1$ has been reduced from 256 to 32.

We can visualize the behavior of a 1-D automaton as a function of time with a **space–time diagram**. Every row of the diagram will describe the automaton's state at a particular generation, with time progressing down the chart. For example, Figure 2.7 shows the progression of rule 18 for an automaton that starts with a single cell with value 1 on a long tape for 20 generations (cells with the value 1 are denoted by a circle and cells with value 0 by a space).

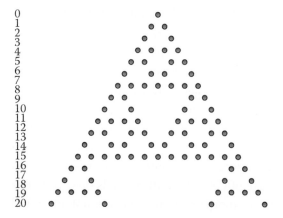

```
0
1
2
3
4
5
6
7
8
9
10
11
12
13
14
15
16
17
18
19
20
```

FIGURE 2.7 Space–Time Diagram. This representation enables one to follow the behavior of a 1-dimensional automaton over time. In this example rule 18 is followed for 20 generations starting from a single non-empty cell.

One of the interesting advantages in the Wolfram notation is that we can go systematically through all automata and qualitatively characterize their behavior. It turns out that roughly 85% of the automata behave in a uniform and uninteresting way (Figure 2.8(a)), 10% in a way that seems ordered and complex (Figure 2.8(b)), and 5% in a seemingly random and chaotic way (Figure 2.8(c)) although they follow deterministic rules. From experiments with other CA it seems that this observation is quite general. In most systems, most sets of rules will lead to uniform and uninteresting behavior (similar to the behavior in Figure 2.8(a)), whereas only in a minority of cases the system will show either a complex behavior (like in Figure 2.8(b)) or random-like behavior (like in Figure 2.8(c)).

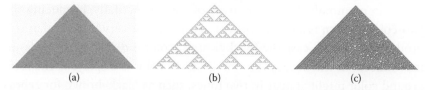

(a) (b) (c)

FIGURE 2.8 The different general behavior of 1-dimensional automata. (a) Trivial order (this specific pattern was created by rule 250); there are 222 automata that show this type of behavior. (b) Complex order, (this specific behavior comes from rule 90); there are 24 such automata. (c) Chaotic-like behavior (created by rule 30); there are 10 automata that show chaotic like behavior among the total of 256 automata. (From Wolfram, Stephen. *A New Kind of Science*, Wolfram Media, 2002. With permission.)

2.5 EXAMPLES OF CELLULAR AUTOMATA

Cellular automata have properties that are useful for modeling certain classes of phenomena. By definition, time progresses discretely for CA, which therefore makes them appropriate for models where time is naturally divided into generations and where the progress from one generation to the next is synchronous. In a cellular automaton, the state of every cell is a function of its neighborhood, and this is appropriate for describing phenomena where the neighborhood affects behavior and the interaction between the components of the model is local. Another important feature of CA is that all the cells obey the same rule, making CA appropriate for modeling homogeneous systems. On the other hand, CA are less useful for modeling systems where time is not discrete or when the behavior of each cell is difficult to define by discrete states.

A few examples will give us a better understanding of the phenomena that can be modeled by CA. Some of these models are based on the previously given CA definition, and in other cases we will modify the CA definition to allow an easier description of the phenomena we are trying to model. Some of the modifications are merely technical; however, some modifications, like introducing randomness into transition rules, are more fundamental.

2.5.1 Fur Color

Mammals' hair color is determined by pigment cells called **melanocytes** in the basal layer of the epidermis (skin). These cells produce **melanin**, which determines the hair color. In general, mammal hair color is a single color or a template containing two colors (e.g., a zebra; unlike fish and birds, no mammals display an array of colors). Actually, producing the two-color pattern is a binary process: either no (or very little) melanin is produced, in which case the "background" color is displayed, or melanin is produced, in which case the "foreground" color is displayed. The foreground color might appear in two tones, such as black–brown for zebras and orange–yellow for tigers. But the main question remains as to how a particular cell knows whether to produce melanin or not.

It is important to realize that this is a developmental decision. Genetically, all the epidermis cells are identical, and there is no a priori determination of what color each cell will display. Therefore, each cell has to determine its melanin-producing behavior based on its neighbors' states. Clearly, the amount of time available for this decision-making process is limited, and at its end each cell has to decide whether to produce melanin. Once this has been determined, the decision cannot be reversed. This process is an example of the more general mystery: the process of **embryonic development**. In general, the embryonic development process depends not only on the organism's genetics but also on many other factors. Understanding the process by which the genetic information is deployed during development in a robust way not affected by irrelevant changes in the environment and resulting in a functioning organism is one of the main challenges of modern biology.

One of the mysteries of this process is that the embryo starts out as a single cell that constitutes a completely homogeneous and symmetrical system. This system develops into a small set of identical cells and then evolves into an asymmetrical system where cells have differentiated such that each may fulfill its unique function (e.g., blood cells differ from skin cells). So the question is how does a homogeneous symmetrical system lose its symmetry? The **symmetry-breaking** question shows up in other scientific fields as well, and although we will not discuss it in detail we will note that Alan Turing, the father of the Turing machine and one of the main figures in the history of computer science, was also puzzled by the symmetry-breaking phenomenon in embryonic development. In 1952 he suggested a theoretical model for this problem (Turing, 1952). This model, just like the model we will present next, is not faithful to the biological details, yet both models show how symmetrical laws of nature that act on a symmetrical system can in principle lead to symmetry breaking. In contrast to Turing's complex model, we will describe two simple models that give rise to symmetry breaking.

Let us start with a very simple model that describes the decision process governing melanin secretion. The cells are located on a grid and can be in one of two colors (e.g., black or white), which are initially chosen randomly. In every generation, each cell chooses one cell in its neighborhood randomly and takes on its color. This process repeats itself for a number of generations. Note that this process does not adhere to our previous CA definition. Here, states are determined stochastically and not by a deterministic rule.

What kind of hair pattern do we expect to see using this model? Obviously, we cannot give an exact answer due to the random behavior of the system. It is important to note that the initial distribution of the black cells is of paramount importance: if the initial number of black cells is very small, the probability that they all disappear after a number of generations increases. In fact, the percentage of cells that start out being black is a *parameter* of the model, and we may change its value and observe its impact on the system's behavior. Once we let the system run for enough generations, we will see the color distribution as spots on the grid; if we limit the neighborhoods to be only the cells to the right and left of every cell or only to the cells above and below it, we will get stripes.

This model demonstrates that we can get complex color distributions by using a simple and local rule for each cell's behavior. This absolutely does not mean that this is the way animals' hair color is determined. When we created this model, we did not take into account what we know about the details of the biological process but concentrated on the abstract problem of symmetry breaking. We could obviously go now and see if the model we came up with corresponds to what is known about the generation of hair color patterns. In this way simple models can be used to guide us toward the relevant properties of the biological system of interest.

On the other hand, the model is too simple to describe the formation of more specific color patterns, such as that of the various tails we see in animals. The limited surface area and the cyclic surface make the color formation on tails more elaborate and specific than the rest of the body.

A slightly more complex model (Young, 1984) considers the weighted average of the cells' colors in a radial neighborhood around each cell. As before, each cell has one of two colors (0 for white, 1 for black), and the transition rule is

$$C'_{i,j} = H\left(\sum_{(i',j')\in N} w(i-i', j-j') C_{i',j'} \right)$$

where C' is the value of the cell in the next generation, w is a weight matrix (note that the matrix is centered around $(0,0)$) used to compute the weighted average, and N describes the radial neighborhood. The function $H(u)$ has value 1 if $u \geq 0$ and 0 if $u < 0$.

Although this system is deterministic, by starting from different random configurations and varying the weights and the neighborhood size (these are the model's parameters) many different color distributions can be obtained. Note that the weights may be either positive or negative. A negative value will result in a phenomenon where a black cell in the cell's neighborhood will prevent or delay the cell turning black.

2.5.2 Ecological Models

CA are extremely useful in simulating ecological systems composed of predators and prey that share a confined habitat.

Consider an automaton with cells in one of eight possible states: 0 for an empty cell, 1 for prey, 2–6 for stages (or ages) in a predator's life, and 7 for a reproducing predator. Predators depend on prey to progress to their next stage, as they die of starvation if there is no prey in their neighborhood. Prey that are adjacent to predators disappear (are "eaten"). The prey reproduce and inhabit adjacent cells in their neighborhoods, but when a predator in state 7 reproduces it fills its adjacent empty cells as well as adjacent cells that have previously contained prey. The cells that have been thus inhabited will contain a young predator in stage 2, while the old parent dies and leaves its cell empty (Ermentrout and Edelstein-Keshet, 1993).

If one starts with a square grid of dimensions 50 × 50 cells, with an initial random distribution of prey, and a few cells containing predators, the system exhibits complex population dynamics. The size of the predators and prey populations fluctuates widely with time. Depending on the specific parameters used, the population can become extinct, show some regular fluctuations, or evolve into a complex prey and predator populations, which interact in unpredictable ways. As we noticed before, most sets of parameters will lead to uninteresting behavior, whereas some will lead to surprisingly rich behavior.

A different ecological model that yields interesting behavior is **Wator** (Dewdney, 1984). In this model, the universe is torus-shaped, filled with water, and contains fish and sharks that hunt and eat them. The fish swim randomly into one of the vacant cells that are adjacent to their cell, either horizontally or vertically. If the fish survives for a few generations and has an adjacent empty cell, it will reproduce into that cell. A shark eats a fish in an adjacent cell or swims into an empty adjacent cell if there are no fish in his neighborhood. A shark that has not eaten for a few generations dies of starvation. The sharks also reproduce in a manner similar to the way the fish reproduce.

This model has five parameters: (1) the initial number of fish; (2) the initial number of sharks; (3) the time it takes fish to reproduce; (4) the time it takes sharks to reproduce; and (5) the time it takes the sharks to starve to death. These parameters can be tweaked so that the system achieves a steady state, but this is a difficult task especially when the game board is small. Typically, this model gives rise to a widely fluctuating system or to the extinction of the population. Note that similar to the color formation automata, Wator uses random moves, and thus the automaton is not deterministic.

It may also seem that Wator is not CA per our definition since the fish and sharks move on the board, while we did not define any motion in a CA and discussed only the way the cells' states change. Actually, this is not a fundamental extension of the original definition. A movement of an organism from one cell to its neighbor can be implemented in the standard model by adding states and rules by which an organism to be moved dies in its original cell and an identical copy is created in a neighboring cell.

2.5.3 Food Chain

This model describes a universe in which different species feed off one another. For example, consider a universe containing grass eaten by zebras, which in turn are hunted by tigers. We will try to model some characteristics of such a food chain using CA.

Every cell in the CA can be in one of N states: $0,1,\ldots,N - 1$. A cell in state k will eat any of its four perpendicular neighbors, which are in state $k - 1$, by changing the neighbor's state to k. The food chain is circular so that cells in state $N - 1$ are eaten by cells in state 0. For example, Figure 2.9 shows a transition of a particular configuration of the CA ($N = 6$, and the states are $0,\ldots,5$). Such automata are called **circular cellular automata** (**CCA**) (Dewdney, 1989). Can you guess how this particular system will behave?

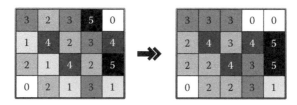

FIGURE 2.9 An example of one generation of the food chain automaton. A cell in a particular state will be "eaten" by a neighboring cell (only the four perpendicular neighbors are considered) with a state higher by 1. The order is cyclic, so a cell in state 5 can be eaten by a cell of state 0. In this automaton spherical boundary conditions are not imposed, and cells on the boundaries have fewer neighbors than interior cells.

2.6 COMPARISON WITH A CONTINUOUS MATHEMATICAL MODEL

For comparison, we present a continuous mathematical model that describes the growth of a bacteria colony in ideal lab conditions. Obviously, this model does not exactly describe the real biological processes.

The reproduction rule we want to implement states that the bacteria population grows at a rate proportional to its size at any point in time. We will denote the number of bacteria at time t by $y(t)$. The rate of growth at time t is

$$\lim_{\Delta t \to 0} \frac{y(t + \Delta t) - y(t)}{\Delta t}$$

In other words, this is the derivative of $y(t)$ which we denote $y'(t)$. So now we can express the reproduction rule as

$$y'(t) = \lambda y(t) \tag{2.1}$$

where λ is a positive constant that characterizes the reproduction rate.

Note that (2.1) describes properties of functions, and we are looking for a function that obeys this equation—in this case, a function whose derivative at any point is equal to the value of the function at that point times a constant λ. Such equations belong to a mathematical field called ordinary differential equations. Solving (2.1) is quite straightforward, as described next.

To determine the function $y(t)$ which gives the number of bacteria at time t, we rewrite (2.1) as

$$y'(t) - \lambda y(t) = 0$$

Multiply the equation by $e^{-\lambda t}$ to get

$$e^{-\lambda t} (y'(t) - \lambda y(t)) = 0 \tag{2.2}$$

The function $e^{-\lambda t}$ is never equal to 0; therefore, a function $y(t)$ will satisfy the reproduction function (2.1) if and only if it satisfies (2.2).

The left-hand side of (2.2) is the derivative of the function $e^{-\lambda t} y(t)$; therefore, (2.2) can be written as

$$\left(e^{-\lambda t} y(t)\right)' = 0$$

The only functions with derivative 0 are constant; therefore, the function $e^{-\lambda t} y(t)$ is a constant, that is,

$$e^{-\lambda t} y(t) = C$$

where C is a constant. To solve for $y(t)$, we divide by $e^{-\lambda t}$ to get

$$y(t) = Ce^{\lambda t} \tag{2.3}$$

This derivation shows that only the exponential functions described in (2.3) satisfy the reproduction rule (2.1); in other words, (2.3) describes the (infinite) set of all solutions to (2.1) and is called the general solution to (2.1).

The reproduction rule described by (2.1) does not provide us with enough information to determine the size of the bacteria colony at time t, as there exists an infinite number of solutions to the equation. This is not surprising since the reproduction rule describes the rate of growth, and the initial number of bacteria has not been specified.

Assume, for example, that the number of bacteria at time $t = 0$ is y_0, that is,

$$y(0) = y_0$$

Substitute into (2.3) to get

$$y(0) = Ce^{\lambda \cdot 0} = y_0$$

Therefore, $C = y_0.$

So of all the possible solutions given by (2.3) only one function satisfies the initial conditions: $y(t) = y_0 e^{\lambda t}$. This function is called the particular solution of equation (2.1) with the initial condition y_0.

This model differs in a few key aspects from the other models we explore in this chapter. In this model time is continuous, as t may be any real number. Furthermore, the mathematical analysis used depended on the change in the number of bacteria over time periods that are arbitrarily short to compute the total number of bacteria. These kinds of models use calculus as their main mathematical tool. The model we just presented is a very simple example: the reproduction rule is an easily solved first-order differential equation. Some models need to use much more complicated differential equations that often are not amenable to analytical solution and require numerical solutions.

The assumptions underlying this model are unrealistic from a biological point of view (e.g., in reality the rate of bacteria reproduction is nonuniform), but they allowed us to focus on the main characteristic of a bacteria colony: that its growth rate is proportional to its size. To build an exact biological model we would have had to also allow for the availability of food, the density of the colony, and so forth. At best, our model describes a bacteria colony during a limited period of its existence. However, using analytical mathematical tools allowed us to derive an exact formula that gives the colony size at any given moment without having to simulate the colony's development during discrete time periods. We can use differential equations to describe more complex processes, for example, processes that deal with multiple organisms competing over resources. This is beyond the scope of this book, but it is important to remember that these tools are available for modeling.

Discrete and continuous models each have their advantages and disadvantages. Some systems are inherently continuous (like blood flow), and some are inherently discrete (e.g., modeling DNA mutations where there are only four types of nucleotides). In many situations, though, the modeler can choose which type of model to use for the task at hand. Often, discrete models are appropriate for populations of fixed size where finite size effects are important, whereas continuous models are more appropriate for analyzing asymptotic behavior. When the equations that govern the continuous model can be solved analytically, the analysis is usually more efficient than the lengthy simulations needed for discrete models. On the other hand, when the equations do not have analytical solution, a discrete model may be preferred.

2.7 COMPUTATIONAL UNIVERSALITY

2.7.1 What Is Universality?

At first glance it seems that the question of whether a computer can be used to solve a particular problem is fundamentally different from the question if an initial state of the "Game of Life" can be found such that a particular behavior may be observed. It would seem that the complexity of the first problem results from the many variables that might affect the answer (e.g., which computer are we referring to, how large is its memory), whereas the analysis of Life does not depend on any such parameters.

Surprisingly, when we analyze what operations computers can perform (i.e., which algorithms can be implemented by them) or even when we try to analyze computing devices that work in diverse and unusual ways, it turns out that in a deep sense all computing devices are able to implement the same class of algorithms. This does not mean that simple computers have the exact capabilities as high-end supercomputers. The supercomputers will arrive at solutions faster, for example, but in spite of the technical differences the fundamental computing capabilities of all computers are the same.

This amazing insight was achieved by researchers who defined different **computational models** and compared them. A computational model is a simple but exact description of the principle characteristics of the operation of a computational device. As we already saw, a model is a simple presentation of a complex system that captures the important characteristics of the system.

The founders of computer science came up with a set of computational models to standardize and formalize the description of algorithms. After carefully defining the models, the researchers attempted to investigate many different questions these models brought up. One such class of questions deals with comparing the different models: determining whether one model is more powerful than another (i.e., given a problem that is solvable by one model, can it always be solved with the other?). Are two models equivalent; that is, are the same sets of problems solvable using both models (even if they might differ in speed or efficiency)?

Studying these problems led to the formulation of a fundamental hypothesis of computer science, the **Church–Turing thesis**. This hypothesis asserts that we can describe formally and precisely the set of problems that may be solved algorithmically. This hypothesis emerged when it turned out that many computational models, which were mostly developed in the 1930s using very different ideas and techniques, were all *equivalent*: any problem that is solvable using one of the models is also solvable in the others. This equivalence led researchers to believe that all the models fundamentally describe the same intuitive idea known nowadays as "effective computation." Any computational model that is equivalent to these models is called a **universal computational model**.

The most famous such model is the **Turing machine** (**TM**). This model, developed in 1936 by **Alan Turing**, is of a state machine that reads an infinite 1-D tape composed of characters. At any point in time the machine is in one of a finite set of states (hence the name state machine), which determines its actions for any input character it reads off the tape. According to the state and the current character on the tape, the machine writes a new character to the tape and moves the read/write head one position to the right or the left of its current position on the tape. The model assumes that any piece of information may be described as a sequence of characters on the machine's tape. A different model was described by the logician **Alonzo Church**. His model is called **Lambda Calculus** and is similar to a minimal computer language and therefore is particularly useful in programming language research.

As already stated, these two models were proven to be computationally equivalent; that is, any algorithm may be translated from one model to the other, and we may always select the model that is easier to work with for proving a particular statement. Nonetheless, the models differ to such an extent that describing a problem in one model might be meaningless in the other; therefore, we need to translate between the models. For example, when describing a computation using a Turing machine we need to describe how the data are encoded on the tape and determine the machine's behavior in every state. But in the lambda calculus there is no tape and no predetermined set of states.

The current state of affairs is that we have a number of models that are computationally equivalent, and no stronger computational model has been found during the decades that have passed since they were first developed. Thus, the current hypothesis in computer science is that a Turing machine can, in principle, perform any computation. Furthermore, Turing proved that some computational problems cannot be solved using a Turing machine. According to the Church–Turing thesis, these problems are algorithmically intractable, and no computer will ever be able to solve them, regardless of any future technological development.

One of the interesting and important characteristics of a Turing machines is that it is relatively easy to build one Turing machine that can simulate the behavior of any other Turing machine. Such a machine is known as a **universal Turing machine**. To build such a universal TM, we need to find a way to describe any other TM on the input tape of the universal TM. This description must contain all the data necessary for a complete description of the TM—that is, its states and the transition rules from state to state upon reading the tape of the simulated TM. Other computational models also allow us to build a universal machine or program that can simulate any other computation under the model.

A universal TM is similar to a modern general purpose computer in the sense of being able to perform a wide range of computations by executing many different programs. Note, however, that in any real computer the memory is finite, as opposed to the idealized infinite tape of a TM.

The fact that we can construct a universal Turing machine, U, is of major importance since this is a single machine with maximal computing power. The existence of universal TMs allows us to replace the question "Does a TM X with the following properties exist?" with "Is there an input for which the universal TM U will behave in the following manner?" Note that the existence of a universal TM implies that for a universal computing model it is impossible to distinguish between the *data* used as input to the computation and the *program* that controls the computational process. The data can be used as the program and vice versa.

Turing showed a well-defined computational problem that cannot be solved by any TM. To understand his counterexample, let us consider a computer program that contains loops. For example, we might implement (in pseudo-code) a program that computes the sum of the integers from 1 to 10 $(1 + 2 + 3...+ 10)$ as

```
sum := 0
for i := 1 to 10
    sum := sum + i
```

At the end of the execution of the loop the variable *sum* will contain the result we wanted to compute. Obviously the statement in the body of the loop is executed exactly 10 times, but for other loops determining the number of iterations that will take place may be much harder. Consider, for example, the following code segment:

```
    sum := 0
L1: if i=0, exit.
    sum := sum+i
    i := i-1
    goto L1
```

Here the situation is more complicated. If the variable i contains the value 10 at the beginning of the execution of the code segment, the loop will be executed exactly 10 times and will calculate exactly the same value as before. If i starts out with any other positive integer value, the loop will be executed i times; however, if i starts out as a negative integer, the condition in statement L1 will never be satisfied, and the computation will be in an infinite loop and will never stop. Since the value of i is determined earlier in the program, it is not clear what its value will be.

It is interesting to ask if we can design an algorithm that will be able to look at a code of a computer program and its input and decide whether it will result in an infinite loop. In simple cases, like the first one we showed, the problem is easy and even trivial to solve. However, for the general case Turing showed that it is impossible to construct a computer program that takes as its input another computer program and its input and determines whether the other program will halt or go into an infinite loop. In other words, it is impossible to determine in a finite time whether a particular program will halt on any particular input. This is called the **halting problem**, and it is unsolvable using our strongest computational model (i.e., in universal computational models). Such problems are called undecidable.

We will see that a cellular automaton is a computational model and will analyze it in comparison with other computational models. Moreover, we will show that the "Game of Life" with its simple rules is a universal computation model. As a corollary, we will see that some questions about the behavior of CA are undecidable and cannot be answered algorithmically.

2.7.2 Cellular Automata as a Computational Model

To discuss CA as a computational model, we first need to explain how the operations of CA can be thought of as performing a computation. Let us look at a particular cellular automaton. At every generation the transition rule of CA determines the state of the cells in the next generation. In other words, we compute for every cell its next state, taking into consideration its current state and the states of the cells in its neighborhood.

If we think, for instance, on the Game of Life, we can rephrase its rules using the following pseudo-code:

```
If the cell is alive
Then
   If the number of live cells in its neighborhood is
   either 2 or 3
   Then
     The cell's state in the next generation is "alive"
   Else
     The cell's state in the next generation is "dead"
If the cell is dead
```

```
Then
    If the number of live cells in its neighborhood is 3
    Then
        The cell's state in the next generation is "alive"
```

It is important to note that the CA executes this computation simultaneously on all cells and therefore that the computation it executes is different from the computation of a single cell's next state. However, a standard computational model (e.g., computer program) can execute this computation cell by cell. Actually, this direction of the proof should be self-evident, as we usually implement CA using a regular computer program. It is harder to show that CA can be used to perform standard computations. To answer this kind of question we have to show how a computation in another model can be simulated by CA.

For instance, let us look at the addition of two numbers. First, we have to represent the numbers in a cellular automaton. Let us consider a 1-D cellular automaton and assume the numbers are represented as unary numbers (i.e., the number n is represented as a sequence of n cells containing 1). The two numbers will be written one after the other, with an empty cell (represented by zero) between them. So an initial state of a cellular automaton

1	1		1	1	1

represents the problem of adding 2 and 3.

Now we need to define the set of states and define the neighborhood so that the CA will eventually achieve the state

1	1	1	1	1	

which represents the number 5.

The problem with finding a transition rule that will apply to any addition problem and not just to the problem 2 + 3 is that the rule has to be applied simultaneously to all the cells of the CA. Moreover, the fixed neighborhood size (in this example it is 3; i.e., each cell sees only its immediate neighbor on each side) makes every cell ignorant of the states of cells outside of its neighborhood. However, in this example we can take advantage on the particular observation that there is only one cell whose local neighborhood is 101, so we can set rules that will be applicable only to cells with this neighborhood. This neighborhood should change to 110. Thus, in the next generation the neighborhood to its right will be 101 and will be treated the same way; after three generations the computation will be correctly executed, no more 101 neighborhood would exist, and thus the computation will end.

More complicated questions can be addressed using CA. Consider a cellular automaton using rule 132 (see Section 2.4), and observe that it can determine whether a number is odd or even. If its initial state is composed of an even number of live cells, then after some generations all the cells will be empty. Conversely, if the initial state is composed of an odd number of live cells, after some generations the automaton will arrive at a steady state where exactly one cell is alive. In other words, this automaton is a special-purpose computer that can determine if a number is odd or even. By adding more states and changing the transition rule, we can build CA that can perform more complex operations, such as squaring any integer and finding prime numbers. The important point here is that we can transform a computational problem so that it can be described by CA, and we can create transition rules that will allow the CA to execute the computation.

In the next section we will see that this can be done in the general case—that is, any computation can be carried out in the CA model.

2.7.3 How to Prove That a CA Is Universal

As already noted, to prove that a computational model is universal we have to prove it is computationally equivalent to another universal computational model such as Turing machines or the lambda calculus. To prove this equivalence we have to prove two claims: (1) that any TM can be simulated by CA; and (2) that any CA can be simulated by a TM. If we proved only one of these two claims, we would have shown that one model is at least as strong as the other and not that the models are equivalent.

The latter direction is simple: while we describe our cellular automata as set of rules we actually run them using a conventional computer program. Since all computer programs can be executed on Turing machines it follows that every automaton can be simulated by a TM.

To prove the other direction—that we can simulate a TM by a cellular automaton—recall that a TM is characterized by the initial state of its tape, the set of states, and the transition rules that determine for every state how the current tape cell has to be modified and how to move the read/write head.

Therefore, we have to determine how to represent the TM's tape and states on the automaton's grid and to write the transition rules such that the automaton mimics the TM's behavior at any point in time. Such a construction is sketched in Section 2.7.4 and proves that we can build a CA for any TM.

Note that this proves only that we can construct a cellular automaton for a given TM. It does not help us to determine for a given automaton whether it is equivalent to any specific TM. Thus, proving, for example, that Life is itself a universal computational model is a bit more complicated, as the transition rules for Life are fixed, and we cannot define the transition rules to simulate the behavior of a particular TM.

The creative solution to this predicament is showing that certain patterns (like the glider described in Figure 2.5) can be made to interact and collectively behave like digital circuits. In particular we can show that we can perform logical AND and NOT operations, which turn out to be sufficient for a universal computation (more about this in Section 2.7.5).

Note that the most general definition of a cellular automaton consists of an infinite number of cells, and therefore the equivalence between a TM and a cellular automaton is relevant only when we limit ourselves to a finite number of nonempty cells in the CA. Remember that a TM performs its computation sequentially and thus cannot access or simulate an infinite number of cells in finite time, whereas a cellular automaton can address an infinite number of cells due to the locality of the transition rule.

2.7.4 Universality of a Two-Dimensional Cellular Automaton—Proof Sketch

We will present a construction of a two-dimensional (2-D) cellular automaton M_s that simulates a given TM M. Showing that any TM can be transformed into a 2-D CA demonstrates that 2-D CA are a universal computational model; the sketch follows the proof in Mitchell (1998).

The automaton M_s simulates a TM M in "real time" (i.e., any step taken by M is simulated in one time unit by M_s). Let M be a TM with n states and an alphabet of size m, and assume without loss of generality that $n < m$.

We note the following properties of M_s:

- The CA M_s operates in an infinite 2-D space. A certain number of its cells will represent the TM and be active, and all other cells are in the quiescent state that will be denoted by 0.

- We will denote by k the number of possible states for each of M_s's cells. Let $k = 1 + max(m,n) = m + 1$. The state numbered $m + 1$ is the quiescent state 0.

- Four cells (h, a, b, s) play a central role in the operation of M_s (Figure 2.10(a)). The cell corresponding to the TM's read/write head is denoted as h and is at state P. The cells adjacent to it to the right and left are denoted as a and b, respectively. The cell currently being read on the tape is denoted by s and is in state S_0. We define the neighborhood of each cell as the seven cells around them arranged in the shape shown in Figure 2.10(b). Note that when each of the four previously mentioned cells is the center of the shape in Figure 2.10(b) it can be identified by unique properties of its neighborhood. For example, the cell s is distinguishable because it has an occupied cell above it and empty cells below it.

The cellular automaton is constructed as follows. One row of the grid is used to represent the tape of M. At each time step t of the operation of M_s the cells in this row are in states corresponding to the characters on the tape of M. Although Turing machines use infinite tape, the size of their input must be finite. In the automaton M_s this is implemented by using a

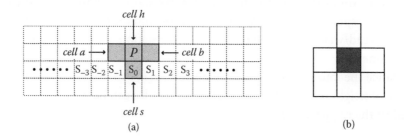

(a)   (b)

FIGURE 2.10   A CA that simulates a TM. (a) A schematic description of the CA's operation. The cell corresponding to the TM's read/write head is denoted as cell h and is at state P. The cells adjacent to it to the right and left are denoted as a and b, respectively. The cell currently being read on the tape is denoted by s and is in state S_0. The other cells depicting nonempty tape cells are denoted by i, and their states are S_i. (b) Description of the neighborhood defined for each cell (marked in gray) in the universal CA. (Adapted from Mitchell, Melanie, In T. Gramss, S. Bornholdt, M. Gross, M. Mitchell, and T. Pellizzari, *Nonstandard Computation*, pp. 95–140, Weinheim: VCH Verlagsgesellschaft, 1998. With permission.)

0	0	0	0	0
0	0	P	0	0
r	t	u	w	y
0	0	0	0	0

➡

0	0	0	0	0
0	0	0	Q	0
r	t	v	w	y
0	0	0	0	0

FIGURE 2.11   The CA before and after executing the following rule: If the head is in state P and sees the input symbol u, change u to v, switch to state Q, and move the head one place to the right.

special state that denotes the leftmost and rightmost cells of the input. All the cells outside of this finite active region are set to the quiescent state 0.

The row above the row representing the tape simulates the read/write head of M, which is denoted as cell h in Figure 2.10. Cell h is exactly above the cell s, whose state S_0 will be read at time t. The cells to the right and left of h are denoted as a and b, respectively. All the cells in this row other than h are in the quiescent state 0. We will not write the transition rules of the CA in a formal way, but we will follow the execution of the CA that simu-lates the following generic rule of the TM M (Figure 2.11): "If the head is in a state P and sees the input symbol u, change u to v, switch to state Q, and move the head one place to the right."

To understand the translation we have to remember that the main problem is that in the CA M_s all the cells act simultaneously according to the transition rules, but to implement M we want to change only the cell that corresponds to the current tape position and the cells that correspond to the read/write head containing the information about the state of the machine.

Since the four previously described cells (s, h, a, b) each sees a different kind of neighborhood (Figure 2.12), the CA rules can be made to uniquely reflect the required changes for each cell.

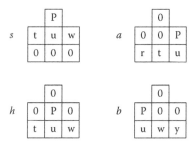

FIGURE 2.12   The unique seven cells environment of the cells s, h, a, b.

Notice that the cell corresponding to the read/write head has to "move" relative to the tape, in accordance with transition table of *M*, but cells do not move in the standard CA model. As usual when designing CA, this problem is solved not by having the cell move but rather by updating the cell values in accordance with the transition rule in a way that simulates information flow between cells. Note that the cell corresponding to the read/write head and the two cells adjacent to it are configured in a way that makes them identifiable by their neighborhoods, and *h* is the only cell among them whose value is not 0. Therefore, it is possible to choose transition rules that update the states of these cells so that the cell describing the read/write head can replace either *a* or *b* (depending on the direction of the head's movement). The state of the new cell representing the read/write head—either *a* or *b*—changes to the state of M_s corresponding to the new state of *M* according to transition table of *M*.

We have thus shown how it is possible to construct a 2-D cellular automaton M_s that simulates any TM *M* thereby proving that 2-D CA are a universal computation model.

2.7.5 Universality of the "Game of Life"—Proof Sketch

The computation universality of Life is proven by showing how to build state configurations that behave as digital circuits that perform logical operations.

We will use the "glider" (shown in Figure 2.5) to build such "digital circuits" composed of logical gates that perform the standard logical operations AND, OR, and NOT.

The rules of Life allow the glider to travel "autonomously" on the board. After four time steps the glider moves one cell diagonally down and to the right. Configurations called **glider guns** "shoot out" new gliders periodically and then return to their initial configuration, which guarantees they will continue to create new gliders indefinitely. Figure 2.13(b) shows the well-known **Gosper glider gun**, named after its discoverer. The Gosper glider gun shoots out a new glider after every 30 time steps (i.e., its period is 30).

The logical circuits are based on using glider guns that shoot out sequences of new gliders representing bits. As an example we will show how to implement the NOT operator. It acts upon a sequence of bits represented by a stream of gliders. The bit 1 is represented by a glider, and the bit 0 is represented by the absence of a glider—a gap in the glider sequence advancing along the board. We place a glider gun that shoots out new

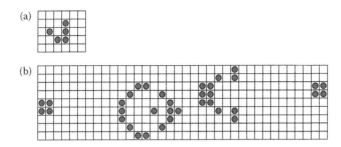

FIGURE 2.13 (a) A glider that can travel autonomously across the board. (b) The Gosper glider gun that releases a glider every 30 generations.

gliders continuously perpendicular to this glider stream. It turns out that two perpendicular gliders colliding in a particular way annihilate each other. Therefore, in the stream of gliders being shot out of the gun only the gliders that do not intersect the gliders representing the input will survive. Put differently, only the gliders that intersect the gaps in the input stream (corresponding to the 0 bits) will survive. Therefore, the surviving stream represents exactly the NOT value of the input stream, as depicted in Figure 2.14.

Notice that, for a glider from the input stream and a glider from the glider gun to annihilate each other, they have to be precisely positioned relative to each other. Different relative positions during a collision can have different outcomes: for instance, a four-cell block or a "blinker" that lights up periodically may be created. Figure 2.15 shows two gliders positioned in an annihilating configuration (verifying this is left as an exercise to the reader).

The operations AND and OR are implemented by employing similar techniques, that is, creating two input streams of gliders (with 1 represented by a glider and 0 by its absence) and letting them interact so that the output stream of gliders will have a glider only if both input streams had a glider (when implementing AND) or when at least one stream had a glider (in the case of OR). To combine these logical operators into more complex logical circuits, which are necessary to show that Life has universal computing power, we need to create more machinery, such as the capability of copying glider streams, moving glider streams, and halting them. We also have to be able to store information (represented as a glider stream) and to retrieve it.

The existence of ways for performing all of these tasks, together with the implementation of the logical operations on glider sequences, shows

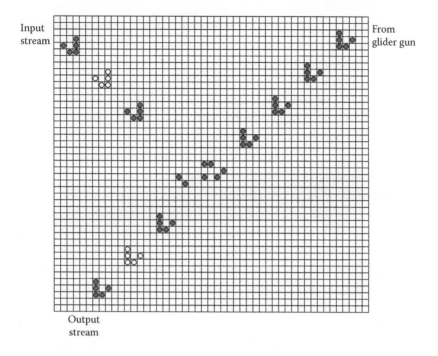

FIGURE 2.14 The configuration of Life that describes the NOT operation on a stream of bits represented by the existence or absence of gliders. Since gliders from the glider gun will survive only if there is a gap (i.e., 0) in the input stream, the output stream will be the logical complement of the input stream.

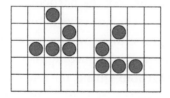

FIGURE 2.15 Two gliders in a configuration that will lead to their mutual elimination.

that it is possible to implement logical circuits that can perform an arbitrary computation within the "Game of Life" framework, which proves that the "Game of Life" is a universal computational model.

2.8 SELF-REPLICATION

Without delving deeply into the problem of defining life, we can say that one of the distinguishing characteristics of life is the ability of living organisms to reproduce. In other words, living organisms can create new organisms that are very similar or even identical to themselves. Recall that von Neumann's purpose when inventing the CA model (together with the famous mathematician Stanislaw Ulam) was to investigate the phenomenon of **self-replication**.

It is easy to find CA that exhibit what looks like self-replication. For instance, a 1-D cellular automaton where a cell with value 1 causes the cells in its neighborhood to also change their values to 1. But this is not an interesting example of self-replication since copying the value 1 from cell to cell is very different from the complex processes of self-replication in the living world which we are trying to investigate, such as cell division, or the creation of a human baby. Therefore, we want to construct a self-replicating cell automaton that is complex enough to convince us that its self-replication mechanism is somewhat similar to replication or procreation in complex biological systems. The cell automaton that copied the value 1 is no more interesting than a rock that splits into two similar-looking smaller rocks during an earthquake, and neither can teach us much about **self-replication.**

Another domain where self-replication can be experimented with is programming, where the challenge is to have a program whose output is the original program itself. This is clearly a programming challenge since **Print** ('**A**') will output **A** and the output of **Print** ('**Print** ('**A**')') is not identical to the original (and in addition might be considered as a syntax error depending on the way the particular programming language handles quotation marks). The recursive nature of the challenge is clear. Surprisingly, such programs were written in almost all programming languages. These programs, nicknamed "quines" after the American philosopher **Willard Van Orman Quine**, operate mainly by tinkering with the printing commands of the particular languages. The following is a Quine program in C:

```
char*f="char*f=%c%s%c;main()
{printf(f,34,f,34,10);}%c";
main(){printf(f,34,f,34,10);}
```

While such programs are fun for programming aficionados, they hardly give us general insight into self-replication.

One way of making the CA models of self-replication more relevant is by looking only at CA that possess a certain minimal level of complexity. This is what von Neumann did. An alternative way would be to look at many CA that exhibit self-replication and then to select those with interesting characteristics. Many researchers chose this path, using models with varying levels of complexity.

The basic difficulty in constructing a self-replicating system is that the system seems to need to contain its own description and the recipe for constructing the next generation, which must contain the recipe for constructing the generation after that, and so on ad infinitum. This is patently impossible.

Von Neumann's insight was that the self-description can have two roles: (1) as a *recipe* for controlling the construction of another copy of the system; and (2) as *data* that will be copied verbatim and attached to the new copy of the system, which in turn will enable it to continue self-replicating.

This approach allows us to solve the infinite regress problem that initially seemed unsolvable. Our current understanding of biology shows that self-replication in living organisms works in a similar fashion, as the DNA that contains the genetic data is read as a recipe for creating proteins and is replicated as data during the replication process. It is amazing that von Neumann had this insight about self-replication already in the 1940s, since the DNA structure was discovered in 1953, and it took a few more years until an understanding of its fundamental properties was achieved.

The self-replicating cellular automaton von Neumann constructed was so complex that he never finished its design completely. The cells could be in one of 29 states, and the automaton had a universal construction capability—that is, it could construct essentially any configuration of cell states based on the description in its input (Mitchell, 1998). Moreover, von Neumann's automaton had universal computing capabilities, that is it can compute any computable function, which as we already saw is highly significant. Further studies were able to complete simpler, but still very complicated, versions of universal self-replicating CA.

It turns out that one can construct relatively simple self-replicating CA if these CA are not required to have universal construction capabilities (i.e., these systems can replicate only specific configurations). **Moshe Sipper** and **James Reggia** (2001) suggested one such elegant system, which was implemented on a 2-D square lattice. Each cell can be in one of five states. It may contain a rook, a bishop, a knight, or a pawn or can be empty. The transition rules for each state are given in Figure 2.16.

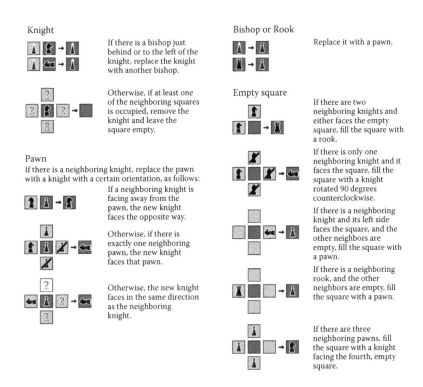

FIGURE 2.16 Rules that describe a self-replicating automaton. Each cell can be in one of five states: rook, bishop, knight, pawn, or empty. (Adapted from Sipper, Moshe and James A. Reggia, *Scientific American* 285, no. 2, 34, 2001. With permission.)

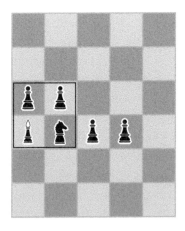

FIGURE 2.17 The initial configuration of the self-replicating automaton. Every configuration in the marked square will be duplicated. (Adapted from Sipper, Moshe and James A. Reggia, *Scientific American* 285, no. 2, 34, 2001. With permission.)

The initial conformation is given in Figure 2.17, where the square at the left is the "genome" and the two pawns to the right form the "replication arm" used in the replication process. We leave it as an exercise to the reader to start from the initial configuration in Figure 2.17 and see how it duplicates. We also leave open the question if such specific self-replicating automata can be used to study the general properties of self-replicating systems.

Note, however, that living organisms are not universal "replicators" either. Bacteria can be used to replicate other foreign or even synthetic DNA, and in fact many applications of biotechnology are based on such capabilities; nonetheless, some genes cannot be replicated in this way as they may be lethal to the host. In higher multicellular organisms there is a much tighter regulation on what can be replicated, and in this sense their replication system is not universal at all.

An interesting alternative way for constructing self-replicating CA is via an evolutionary process. Here we start with a large collection of randomly created CA and then search for the ones closest to being self-replicating, that is, the CA that create copies closest to their initial configuration. At the next stage, one starts with the previously identified set, mutates them slightly, and repeats the process. By iterating this process we may eventually find CA that replicate exactly. This search process is similar to biological natural selection, which is the engine of the evolutionary process. We will deal with this search procedure in more detail when we discuss genetic algorithms (Chapter 3) and Avida (Chapter 6), a programming environment that enables experiments related to artificial life and can be used, among other purposes, to study self-replication.

2.9 SUMMARY

We have seen that CA originated in an effort to model biological processes (e.g., cells, bacteria colonies), but they are useful in other contexts as well.

Models based on CA also abound in physics, chemistry, and other areas in which there is a need to build models based on discrete time and space. An interesting application is discussed in Rosin (2006) where cellular automata were trained to perform image processing tasks like noise filtering, thinning, and finding convex hulls. The purpose of the training was aimed to automatically select, using a technique similar to genetic algorithms, a set of rules that can perform the task at hand. CA are also useful where research on the origin of complex behavior from simple local rules is being done.

The main lesson of this chapter is that a small number of simple rules repeatedly applied can create a wide spectrum of complex behavior that often seems as if it were the result of a detailed plan and nonlocal coordination.

Considering CA as a nonstandard computational model enabled us to use fundamental concepts and theorems from computer science to gain deeper insights into CA. One such aspect is our study of universality. We should note that universality is interesting as a theoretical question that determines the theoretical capabilities of cellular automata. Models of specific phenomena do not depend on universality, and, furthermore, no one would use a cellular automaton as a practical universal computational device.

In our theoretical studies of CA we were able to use the undecidability of the Halting problem to learn that the fate of initial configurations in the "Game of Life" cannot be predicted without actually following the simulation all the way through. In more grandiose terms, Life cannot be predicted; it must be lived!

2.10 PSEUDO-CODE

```
// Generic code to run a 2-D cellular automata

// Initializing the matrix to the starting set-up
INIT_MAT(mat)

WHILE not END_CONDITION(mat)
  BEGIN
  // The end condition can be met by either reaching a pre-determined
  // number of generations, or by reaching a certain state of the matrix.


      FOR i:=1 TO n
       FOR j:=1 TO n
         // Calculate the new state of each cell based on its current
         // state, the states of its neighbors, and the transition rules
         // which constitute the function NEW_STATE.
         // To emulate simultaneous update of the main matrix a temporary
         // matrix is used.

         temp_mat[i,j] := NEW_STATE(NEIGHBORHOOD(mat,i,j))

       // Update the main matrix

      FOR i:=1 TO n
       FOR j:=1 TO n
        mat[i,j] := temp_mat[i,j]
  END
```

2.11 FURTHER READING

Ben-Jacob, Eshel. 2008. Social behavior of bacteria: From physics to complex organization. *European Physical Journal B-Condensed Matter and Complex Systems* 65, no. 3, 315–322.

Dewdney, Alexander K. 1984. Sharks and fish wage an ecological war on the toroidal planet Wa-Tor. *Scientific American* 251, no. 6, 14–22. See implementation at: http://wator.panmental.de.

Dewdney, Alexander K. 1989. Computer recreations: a cellular universe of debris, defects and demons, *Scientific American* 261 no. 2, 102–105.

Ermentrout, G. Bard, and Leah Edelstein-Keshet. 1993. Cellular automata approaches to biological modeling. *Journal of Theoretical Biology,* 160, no. 1, 97–133.

Mitchell, Melanie. 1998. Computation in cellular automata: a selected review. In T. Gramss, S. Bornholdt, M. Gross, M. Mitchell, and T. Pellizzari, *Nonstandard Computation,* pp. 95–140. Weinheim: VCH Verlagsgesellschaft.

Rosin, Paul L. 2006. Training cellular automata for image processing. *IEEE Transactions on Image Processing* 15, no. 7, 2076–2087.

Shapiro, James Allen and Martin Dworkin (Eds.). 1997. *Bacteria as Multicellular Organisms.* Oxford: Oxford University Press.

Sipper, Moshe and James A. Reggia. 2001. Go forth and replicate. *Scientific American* 285, no. 2, 34.

Turing, Alan M. 1952. The chemical basis of morphogenesis. *Philosophical Transactions of the Royal Society B* (London), 237, 37–72.

Wolfram, Stephen. 2002. A new kind of science, *Wolfram Media.* Available at: http://www.wolframscience.com/

Young, David A. 1984. A local activator-inhibitor model of vertebrate skin patterns. *Mathematical Biosciences* 72, no. 1, 51–58.

2.12 EXERCISES

2.12.1 "Game of Life"

1. Given the initial state of the board in Figure 2.18, compute the state of the board in the next five generations. Assume the grid is large enough not to encounter boundary conditions.

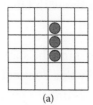 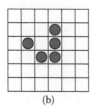 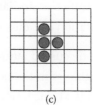

(a)   (b)   (c)

FIGURE 2.18

2. Why did we need pieces in two colors for the manual execution of Life as described in the text?

3. Does the orientation of the live cells on the board affect the system's outcome? That is, if, for example, we rotated the boards in exercise 1 by 90° clockwise, would the future of the system change? How?

4. Try to find an initial configuration for Life that will result in an oscillator with period 3 (i.e., returns to its initial state after three generations).

5. True or false: If the board is finite, one would eventually return to states that have occurred previously in a periodic fashion. Prove your claim.

2.12.2 Cellular Automata

6. In the standard CA model, how would you build an automaton where every cell at every instant belongs to one of two types of cells, for which there are different transition rules (the neighborhoods are the same for both types)?

7. There are 256 one-dimensional automata with $k = 2$ and $r = 1$. Find a formula for computing the number of automata as a function of the parameters k and r.

8. Describe rule 146 using a table.

9. Choose a random starting position, and create a space–time diagram 10 generations long for executing rule 146 from this initial state.

10. Identify the similarity in behavior between rules with binary representation $a_1 a_2 a_1 a_2 a_2 0 a_2 0$ (rules 0, 90, 160, 250).

11. We discussed automata where cells can determine the state of their neighbors in the next generation. Explain why this does not contradict the basic model where each cell determines only its own state in the next generation.

12. Explain how it is possible to simulate a CA with any neighborhood using a CA with a nearest neighbor's neighborhood.

13. Given a deterministic CA operating on a finite board, prove the following: If it has a configuration that can be reached in one step from two previous configurations, a configuration of the CA can be found that has no previous configuration. Such a configuration can therefore exist only as an initial configuration of the system and is called a "Garden of Eden" configuration.

14. It is possible to change the CA model to a model with asynchronous time by randomly choosing only one cell at each time step and applying the transition rule only to that cell. Does this change the properties of the model significantly?

15. We have defined the transition rule in a way that is independent of the cell's location on the board. What is the importance and the goal of this property? What would happen if we removed it?

16. Construct a 1-dimensional CA with three states (0,1,2). The initial state of the automaton is composed of a pair of 1s, and all the other cells are in the quiescent state (state 0). See the space–time diagram in Figure 2.19. The transition rule should change the states of the cells between the two 1s from left to right from state 0 one after the other to state 2 over time. After changing the rightmost 0 to 2, the CA will not change anymore. Ignore what happens outside the boundaries marked by the two 1s.

1	0	0	0	1
1	2	0	0	1
1	2	2	0	1
1	2	2	2	1

FIGURE 2.19

17. Construct a 1-dimensional CA using Exercise 16 as a reference point; however, state 2 will now progress from left to right between the 1s, with the provision that at any generation only one cell will be in state 2 (Figure 2.20), and state 2 will disappear in the last generation. Hint: You may need to add extra states to the automaton.

1	0	0	0	1
1	2	0	0	1
1	0	2	0	1
1	0	0	2	1
1	0	0	0	1

FIGURE 2.20

18. Construct a 1-dimensional CA using Exercise 17 as a reference point; however, state 2 should now move right and left indefinitely. That is, when the automaton reaches the final state of Exercise 17, it will then make state 2 move left until it will meet the cell in state 1 on the left and then will turn right again, thereby creating an indefinite oscillator.

2.12.3 Computing Using Cellular Automata

19. Why could we discuss the undecidability of the halting problem for computer programs and Turing machines as if it was one claim, even though the two computational models are different?

20. The following problem is undecidable: Given a computer program and an input string, determine whether the program will output a specific character during its execution. Show how we may use the halting problem to prove the undecidablility of this problem.

21. List the main differences between a computational model based on Turing machines and a model based on cellular automata.

22. We have claimed that we can construct a two-dimensional CA that simulates any TM. Is this enough to prove that CAs are a universal computational model, or do we need to apply the construction for a universal TM and find a CA equivalent to it?

23. We have claimed that we can construct a CA for a given TM. From this claim we can deduce that we can construct a CA that can accept on its grid a description of a TM and can simulate it. Why?

2.12.4 Self-Replication

24. Von Neumann's automaton has a universal construction capability. What input does it need to construct a self-replicating automaton?

25. Based on the results given in Section 2.8, is a universal construction capability necessary for self-replication? Is it enough?

2.12.5 Programming Exercises

26. Program an "engine" to execute any CA. When designing your program try to allow maximum flexibility in defining the CA's universe (the grid on which it computes) and the transition rules.

27. Using the engine created in Exercise 26, implement the automata mentioned in this chapter.

28. Construct a cellular automaton that, when given a board containing cells that form all kinds of shapes, will leave only shapes that look like a symmetric cross of any size (i.e., lines of the same size that cross each other in the middle) and will eliminate all other shapes (see Figure 2.21). (A single cell is not considered a cross.)

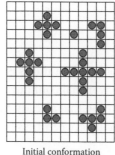 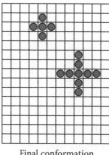

Initial conformation   Final conformation

FIGURE 2.21

29. An $N \times N$ board where each cell has a random numerical value between 0 and N^2 is given (assume that N is known to the cells). Find the rules for a cellular automaton that will keep only the cells with the highest value on the board and will eliminate (set to 0) the values of all other cells (see Figure 2.22).

0	6	4	2
3	12	9	9
5	8	7	5
3	4	12	0

Before

0	0	0	0
0	12	0	0
0	0	0	0
0	0	12	0

After

FIGURE 2.22

2.13 ANSWERS TO SELECTED EXERCISES

4. Many patterns can be found with three generations periodicity. One such example is given in Figure 2.23.

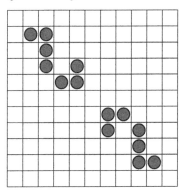

FIGURE 2.23

5. If the board is finite, the number of configurations is finite. (For an $N \times N$ board it is $2^{(N \times N)}$). Thus, for any initial configuration, its evolution must eventually hit a configuration that already occurred. Since the rules are deterministic, the cycle between these configurations will repeat itself forever.

7. The number of possible automata is $K^{k^{(2r+1)}}$.

10. In all of these rules the state of a given cell in the next generation is dependent on the state of its two neighboring cells but not on its own state.

13. Since the automaton is deterministic, every configuration is followed by a unique consequent configuration. If there is a configuration that is reachable from two different previous configurations, it means that the number of consequent configurations is smaller than the total number of configurations of the system. Therefore, the system must have some configurations that are not the consequents of any configuration and are thus "Garden of Eden" configurations.

15. The requirement that a cell cannot use its location (e.g., its x,y coordinates) in calculating its next state is essential to the concept of cellular automata as it allows each cell to communicate only with its local neighborhood and requires that all cells will follow the same

rules. It thus ensures homogeneity and locality and ensures that the computation of the CA is not a simple translation of programs that use addressable memory.

16. The key here is to notice that the new state of the cell is determined by its current state and the current state of its neighbor from the left according to the following table. Note that there are other combinations of the two relevant bits (i.e., the current cell and the cells to the left) but they will not occur inside the boundaries marked by the two 1s.

Left	Current	New State
0	0	0
1	0	2
2	0	2
0	1	1
2	1	1
2	2	2
1	2	2

17. The solution requires changing the current 2 into 0 and changing the 0 to the right into 2. However, the leftmost 0 cell must use an additional flag to prevent it from starting another wave of 2 after the first one.

20. Add a print command for the required character immediately before any stop command (or termination point) of the program. (If that character is part of any other output of the program replace the character for that output.) Now, if this printing problem was decidable, then we would have a solution to the Halting problem, which is known to be undecidable and, therefore, this is impossible.

24. To construct a self-replicating automaton the input tape should contain a coded description of the automaton.

CHAPTER 3

Evolutionary Computation

3.1 EVOLUTIONARY BIOLOGY AND EVOLUTIONARY COMPUTATION

3.1.1 Natural Selection

Charles Darwin opened his famous book, *The Origin of Species,* in which he presented the theory of evolution by **natural selection**, with a discussion aimed at showing that a far-reaching hereditary change in organisms' characters is possible and that such changes can be achieved by selective breeding. Darwin explained the mechanism for such evolutionary changes by presenting the example of domesticating animals and crops. By selective breeding based on small variations in hereditary characteristics, one can gradually create different strains according to the breeder's preferences. Figure 3.1 gives as an example *Brassica oleracea* and the variety of crops cultivated from it. Domestication and **artificial selection** often result in extensive hereditary changes. Consider, for example, the different dog breeds or the many edible and decorative plants created by man. Domestication is an example of evolution in action, albeit on a small scale. Darwin argued that under natural conditions the environment takes on the role of the breeder, as individuals who are better adapted to the environment reproduce more than others.

The industrial melanism of the **peppered moth** provides a famous example of this process. An increase in air pollution has been shown to give a camouflage advantage to darker moths that were rare before the industrial revolution, so that the dark form of the moth became the more prevalent form, replacing the previously more common light form. Antipollution

FIGURE 3.1   Some of the cultivated varieties of wild cabbage (*Brassica oleracea*).

legislation, which led to reduced pollution, reversed the trend. Scientists have observed a more disturbing example of evolutionary change: various strains of bacteria have developed resistance to antibiotics, often very rapidly.

The theory of natural selection has far reaching implications, but it is based on simple basic assumptions:

1. There is variation between the individuals in the population, each individual having a unique combination of characteristics.

2. A large part of this variation is hereditary.

3. The world has limited resources, and some of the variants in the population can make better use of these resources. These individuals will produce more offspring than other individuals.

These simple and widely accepted assumptions necessarily lead to natural selection. In other words, the principle of natural selection is an unavoidable result of the conditions that exist in the living world.

One of biggest obstacles to Darwin's natural selection theory was understanding the source and nature of the **hereditary variations**. When Darwin published *The Origin of Species* in 1859 there was no established theory explaining heredity. It is interesting to note that in 1866 **Gregor Mendel**, the father of modern genetics, discovered the basis for hereditary laws, but his research was ignored and forgotten until the beginning of the 20th century. At the beginning of the 21st century, we now have a much more complete picture of the mechanisms of heredity based on our understanding of DNA and the processes related to it, as we saw in Chapter 1. Darwin assumed **blending inheritance**, in which the characteristics of both parents blend to produce the corresponding organ in the offspring. Darwin believed that tiny particles he called **gemules** allow for heredity. The gemules represent characteristics, are present in all organs, and are sent from the organs to the reproductive cells so that during reproduction the gemules from both parents mix together to create the offspring, who usually has the average of the parents' characteristics. In turn, the offspring's gemules will be sent to their reproductive cells, and so on.

Darwin's blending inheritance theory, also known as **pangenesis,** had severe theoretical problems since selection (natural or artificial) cannot affect blended characteristics. Repeated blending will "dilute" the characteristics, and selection cannot preserve them. In the same way that when we mix red and white paints we cannot separate the resulting pink back into its white and red components by more mixing, it is impossible to separate the parents' characteristics after blending. As discussed in Chapter 1, we know now that the alleles of both parents are preserved in the offspring and are not blended. This was one of Mendel's most important findings.

Another important problem with Darwin's inheritance model is that it allows the inheritance of acquired characteristics, which the individual developed during his or her life. If an ostrich developed calluses on its knee due to the knee rubbing the ground when it runs and passed this property on to its descendants so that they will have calluses on their knees even before they start running, this would be the inheritance of acquired characteristics. Since young ostriches really do have calluses before running, this idea seems plausible, but we know now that ostriches have calluses on their knees at a young age because of natural selection and the reproductive advantage of individuals with this property and not due to the inheritance of acquired characteristics. Or consider the long necks of giraffes. According to the inheritance of acquired characteristics model, ancient giraffes with necks of normal length were compelled to feed on leaves of tall trees and thus had to stretch their necks. This acquired trait was passed on to their offspring, resulting in the long neck of modern giraffes. According to pangenesis when a particular organ is affected by the environment, its gemules are affected in the same fashion and are passed to the reproductive cells in this new form and therefore the acquired characteristics are passed on to the individual's descendants. Inheriting acquired characteristics is often referred to as **Lamarckian evolution**, named after **Jean Baptiste Lamarck**. He predated Darwin and suggested an evolutionary model that relied on the inheritance of acquired characteristics.

Note that there are two types of information flow according to this model of inheritance of acquired characteristics. First, the change caused by the environment (e.g., the callus caused by the knee hitting the ground) has to be stored as information in the gemules. As the reproductive cells do not contain a copy or "image" of the organs but rather instructions that control the gradual and complex development of the embryo and the individual under the environmental influences, it is unclear how to automatically translate back an external change in an organ to a change in these instructions. **August Weismann**, one of the most influential scientists to have worked on evolution, described this problem as follows: believing that information can be translated in such a manner is similar to believing that an English telegram sent to China will arrive already translated to Chinese. The second information transfer, according to the model of the inheritance of acquired characteristics, is the passing of the gemules from the organs to the reproductive cells. This would require that reproductive cells do not exist early on but rather are produced during the organism's adult life from the gemules sent from the various organs. This information transfer is not supported by modern understanding of the major hereditary processes. For example, we know now that all human female eggs, or oocytes (about 500,000) already exist when a baby female is born. Male sperm cells are being continuously produced in the testis, but we don't know of a mechanism that enables transfer of arbitrary information from the rest of the body into the testes. Weismann actually tried to check the validity of the model by repeatedly breeding mice whose tails have been cut off and discovered (of course) that the offspring were born with tails of regular length.

Interestingly, recent discoveries may open the door to the return of Lamarckian ideas (of course not in their naïve version), operating at the molecular level. Mechanisms like reverse transcription, which enables reverse flow of information from RNA to DNA in contrast to the central dogma (see Section 1.3.4), and inherited regulatory changes in gene *expression* (a form of inheritance referred to as **epigenetic inheritance**) paint a richer picture of hereditary information flow than was previously assumed. In any case, even though Lamarckian evolution differs from our modern genetic model, it can be used in evolutionary computation (genetic algorithms, or GAs). We will see in Section 3.5 how this idea is implemented and to what effect.

Modern evolutionary theory (known as **neo-Darwinism**) denies the troublesome ideas of blending inheritance and the inheritance of acquired characteristics. It is intimately related to the modern understanding of heredity provided by genetics, based on distinguishing between inherited genetic information and the organism's characteristics. The transfer of genetic information is schematically described in Figure 3.2. The individual's genetic information is called its **genotype** (multiple units of inheritance called *genes* make up the genotype). The individual's expressed characteristics are called its **phenotype**. For example, the information in the DNA molecules (see Chapter 1) that is responsible for eye color is part of the genotype, whereas the eye color itself is a phenotype. Hereditary changes impact the genotype, whereas natural selection selects between the different phenotypes created under the combined influence of the genotype and the environment. We return to the distinction between genotype and phenotype in the discussion of the use of Lamarckian inheritance by genetic algorithms.

The origin of variations in the genotype is random mutations. In **asexual reproduction** a single parent passes a copy of all his or her genetic information to the offspring. Therefore, each descendant is an exact genetic copy of the parent, except for rare **mutations**. On the other hand, in **sexual reproduction** each individual usually contains two copies of genetic information—one from each parent—which are not necessarily identical, since each parent may pass on a different form of the gene. The different forms of a gene are called **alleles**. Additionally, during meiosis new combinations of alleles taken from the two parental chromosomes are created (see Chapter 1). Therefore, in sexual reproduction, the progeny have new combinations of alleles and thereby different characteristics. This is the source of the power of sexual reproduction, as found in animals and humans. Asexual reproduction is found in bacteria, plants (which exhibit both types of reproduction), and some other simple organisms.

To summarize, neo-Darwinism, which is the common evolutionary theory held today, denies the passing on of acquired characteristics and explains

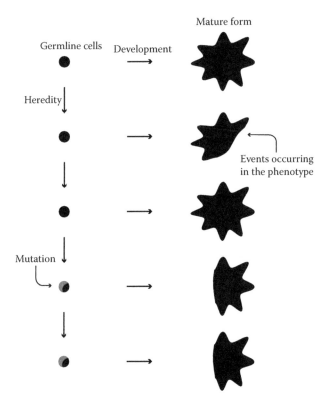

FIGURE 3.2 Information transfer in asexual reproduction. The hereditary continuity is based on reproductive (germ line) cells. The information in the reproductive cells is transferred to the next generation. Only changes in the genetic makeup of the reproductive cells (i.e., mutations, which are rare) impact the next generation. Changes to the nonreproductive cells of the organism (somatic cells), which are caused by the environment, will change the mature form but will not impact the offspring. The information transfer in multicellular organisms with sexual reproduction is also mediated by germ line cells, but the situation is more complicated, as each offspring is endowed with a unique combination of genes from its parents.

evolutionary change by natural selection of nondirectional (random) variations. For our discussion it is important to emphasize the following points:

1. Genetic change can result from one of two events: A *mutation*, which is a discrete random change in the genetic material that causes a change in a trait (e.g., color change), or, in sexually reproducing organisms, via a *new combination* of alleles.

2. Natural selection is not a creative force in evolution; it can eliminate deleterious variations in the population and can increase the frequency of successful variations.

3.1.2 Evolutionary Computation

The term **genetic algorithm** in its most general sense refers to a family of computational models inspired by biological evolution. The different models are based on various conceptions of the biological evolutionary processes and do not necessarily faithfully represent current biological understanding. Moreover, genetic algorithms allow researchers to experiment with different evolutionary mechanisms in order to analyze their behaviors and outcomes.

Any evolutionary computation is based on representing possible solutions to a computational problem as "genetic" information passed from one generation to the next. The evolutionary process is initialized with a (usually random) **population** of "solutions," which correspond to individual organisms. Each solution is examined to determine how successful it is in solving the computational problem; this defines its **fitness** and determines how well represented it will be in the next generation's population. Furthermore, new solutions may be created by introducing random changes (similar to *mutations*) or by combining elements of different solutions (similar to *sexual reproduction*). Therefore, better solutions become more prevalent from one generation to the next, and the probability of finding a satisfactory solution to the computational problem increases. The computational process is halted when a good enough solution to the problem is found. Evolutionary computation is often described as a function optimization process, since the computation can be viewed as a search for the maximal value of the fitness function.

Genetic algorithms are used for a variety of applications, prominently optimization problems. Typically the problems are hard to attack using standard mathematical tools (e.g., the functions are noncontinuous, the set of equations is nonlinear), and the set of candidate solutions is very large. An important advantage of genetic algorithms is that it is a very generic method, and a genetic algorithms engine can be utilized for solving a wide variety of different types of problems by modifying or supplying a few functions. A good review of evolutionary computation can be found in Mitchell and Taylor (1999).

3.2 GENETIC ALGORITHMS

Solving a computational problem using an evolutionary process depends on answers to a few fundamental questions:

1. How is the genetic data passed from generation to generation represented?

2. How is the fitness of each individual determined?

3. How does fitness affect the makeup of the next generation?

4. Which genetic changes will be performed on individuals when producing the next generation?

5. When should the evolutionary computation terminate?

The classical genetic algorithms model, which we will use during most of this chapter, is based on the model first suggested by **John Holland** in 1975 (Holland, 1975). In this model, the answer to the first question is that the genetic information is represented as a fixed-length sequence of binary bits. We refer to every such sequence as a **genotype** or sometimes as a **chromosome**. It is important to understand that, while this particular representation is the most familiar one, there is nothing sacred about this representation. Other representations such as a sequence of numbers rather than a sequence of bits, arrays, matrices, linked lists, or other data structures may be used as representations when they are more suitable. The important thing is to preserve the principle that random changes in the solutions will change their fitness, which in turn will change the impact of each solution on the population of solutions in the next generation of solutions, and with that will give rise to the possibility of generating better and better solutions.

In the algorithm we will present, the offspring are generated by combining the chromosomes of two parents. The offspring, which like all individuals are made up of a single chromosome (i.e., the organisms are "haploids" and carry only one chromosome), contain genetic information derived from both parents by a **crossover** process similar to the biological process of exchanging genetic information between chromosomes.

The crossover process of two chromosomes is done by randomly picking a location in the sequences and exchanging the corresponding parts of both chromosomes:

```
10110 \/ 0101011

01010 /\ 1100100
```

After exchanging the corresponding parts, the following two chromo-somes will be generated:

```
101101100100
010100101011
```

The general structure of a genetic algorithm is as follows:

1. Start with a random population of n chromosomes.

2. Compute the fitness of each chromosome in the population.

3. Repeat until n offspring are created in the new population:

   a. The **selection** phase: Pick a random *pair* of chromosomes from the current population. The probability of picking a particular chromosome has to be an increasing function of fitness.

   b. The **crossover** phase: Crossover the parent chromosomes with probability p_c (the **crossover probability**), and then choose arbi-trarily one of the resulting chromosomes as the offspring (if there was no crossover, choose one of the parents).

   c. The **mutation** phase: For every bit in the offspring, flip it (if 0 then change to 1; if 1 then change to 0) with probability p_m (the **mutation probability**).

   d. Insert the offspring into the new population (note that the pro-cess in steps a through c generated a single offspring).

4. The new population becomes the current population forming the next generation.

5. If a predefined end criterion has been reached then stop; otherwise repeat from step 2.

There are many valid variants within this scheme, some of which will be discussed later in this chapter. The general description shows that the behavior of a genetic algorithm is influenced by a few central param-eters. The general structure of the algorithm does not dictate how the fitness of each individual is computed and how the differences in fitness determine how the parents, who will mate and create the next generation, are chosen. These two components, as well as the probabilities p_c and p_m, have a major influence on the behavior of the algorithm, that is, on the changes in the population from one generation to the next. Another important parameter is the population size n, which determines how many individuals will be tested in each generation out of all the possible solutions or chromosomes.

EXAMPLE

Let us demonstrate how a genetic algorithm behaves using a very simple example where the goal is to find the integer x in the range $[0,...,31]$ that maximizes the function $g(x) = 31x - x^2$.

Clearly, this is a very simple problem that is easily solved analytically (or by exhaustive search), but it will serve to demonstrate how to solve a problem using a genetic algorithm (Table 3.1a and Table 3.1b).

Solution representation: we will represent every possible solution by a five-bit sequence (chromosome). Five bits suffice exactly to represent the integers 0 (00000) to 31 (11111) as binary numbers.

Initial population: we will choose the population size to be 4 (just for the sake of the example, as this population is much smaller than commonly used). The initial population will be random.

Score: the score of an individual i in the population, f_i, will be the value of the function g; that is, we will look at each chromosome as the binary representation of a number x and will compute $g(x)$.

The probability of selecting a chromosome as a parent: the probability of picking individual i as a parent is correlated to its score and is defined as the ratio between its score and the sum of the scores of all organisms in the population, that is,

$$P(i) = \frac{f_i}{\sum_{\text{all solutions } k} f_k}$$

Sample run: Table 3.1 shows the execution of the algorithm for a population of four individuals.

- The sequences 10011 and 11000 are chosen as parents. For the crossover phase, the point after the fourth bit was randomly picked as the crossover point to produce 10010 and 11001. We pick the first chromo-

TABLE 3.1a   Initial Generation of Genetic Algorithm Optimizing g (see text for details)

Initial population	10011	01000	11000	01101
	($x = 19$)	($x = 8$)	($x = 24$)	($x = 13$)
Score	228	184	168	234
Probability of being selected as parent (rounded)	0.28	0.23	0.20	0.29

TABLE 3.1b   Population of the Second Generation of Genetic Algorithm Optimizing g (see text for details)

Next generation	10010	01101	11001	01011
	($x = 18$)	($x = 13$)	($x = 25$)	($x = 11$)
Score	234	234	150	220
Probability of being selected as parent (rounded)	0.29	0.29	0.18	0.27

some and select it as the first individual in the next generation (note that we are not demonstrating mutations in this example).

- This process is repeated until we have the next generation, that is, the four offspring shown in Table 3.1b.
- The entire process (i.e., score calculation, selection, and crossover) is repeated to produce subsequent generations.

When will the execution of the algorithm halt? We can choose among a few strategies, depending on the problem:

- If we know what the score of the optimal solution should be, we can halt when it is reached.
- If we know what the score of the optimal solution is, we can halt when an individual with sufficient score has been found (e.g., 90% of the optimal).
- We can halt when the variance in scores in the population is small (a special case is to halt when all the individuals have the same score).
- We can decide to halt if the fitness of the best solution found has not improved for a certain number of generations.
- We can decide to halt after a certain number of generations (e.g., halt after 1000 generations) and then choose an individual with the maximal score as the solution. Since it is in general possible that the computation will not converge to an optimal solution, this technique can be combined with any of the others to ensure termination.

Based on prior knowledge of $g(x)$ we might choose to halt the evolutionary process once an x with $g(x) \geq 240$ is reached or at most after 1000 generations.

It is probable that after some generations the individuals 10000 or 01111 which have the score 240 will be generated and the system will halt (in the range [0,…,31], $g(x)$ reaches a maximum value of 240.25 for $x = 15.5$; however, in this example we considered only integer values for x).

The beauty and strength of the genetic algorithm is that we did not need to understand and analyze the function g, so we could use exactly the same algorithm to maximize other, much more complex functions, for example, $g(x) = 15 - (x - 2)^2$ or $g(x) = \sin(x)^3 - \cos(x)$.

3.2.1 Selection and Fitness

The phrase *survival of the fittest* is often used to describe the process of natural selection. This term, which was first used by Herbert Spencer and not by Darwin, means that, on average, the individuals best adapted to their environment survive and reproduce the most in nature. For example, animals that make good use of their food resources or that can escape from their predators by running fast are adapted to their environment and have increased chances of survival. Assigning a precise meaning to the phrase survival of the fittest is elusive, however, as it seems to be making a circular claim: survival is the result of being adapted, while being adapted is defined as survival in the environment. According to this interpretation it would seem that survival of the fittest boils down to survival of those that survive. Biologists have argued frequently about the meaning of the claim that evolution works by the principle of survival of the fittest. Most biologists today prefer to use the original term natural selection to describe the evolutionary processes.

Natural selection of course does not compute a numerical fitness value for each individual that in turn determines how many descendants he or she will have. It is the other way around: the number of offspring defines the fitness of the individual. In general, if an organism has many offspring we can deduce that it is better adapted.

Genetic algorithms are obviously very different, as the fitness function (called the score in the previously given example) chosen by the programmer has a direct impact on the number of offspring. The choice of fitness function can cause the algorithm never to converge to a solution or to do so very slowly. The values of the fitness function are the inputs of the selection process, so the fitness function and the selection method have to be compatible for the genetic algorithm to perform successfully.

It is important to distinguish between computing a measure of the success of a *particular* solution in the population (a single chromosome) and the *relative success* of a solution in relation to other solutions in the population. In the previous example, f_i was computed without taking the other chromosomes in the population into account, whereas the values $P(i)$ were computed by taking all the scores of the population into account. To capture this distinction we introduce the following definitions:

- **Evaluation Function**: a function used to determine the success (score) of a single solution, based on the requirements of the problem.

- **Fitness Function**: a function which translates the value of the evaluation function to the value which will determine how likely the solution is to reproduce (that is, how frequently it will participate in the reproduction and crossover operations).

The values of the evaluation function for each individual in the population are independent of each other. On the other hand, the value of the fitness function is always defined relative to a given population (usually relative to the values of the evaluation function for all other individuals in the current generation).

The fitness of individual i is usually defined to be $f_i / \bar{f}$ where f_i is the value of the evaluation function and $\bar{f}$ is its average value over all the individuals in the population. This normalization allows us to consider the relative quality of individuals in the population when selecting the individuals that serve as parents of the next generation.

Recall that after computing the fitness values, the next phase is the selection phase in which individuals are selected as parents according to their fitness. Before discussing various selection regimes, we define a term that will help us analyze the difference between different selection mechanisms.

The **selection pressure** is the degree by which the genetic algorithm prefers selecting individuals with a high-fitness value as parents for the next generation over individuals with an average or low-fitness values. Intuitively, a selection method with high selection pressure creates more copies of the better individuals, thereby hastening the removal of the individuals with lower fitness values. In other words, when the selection pressure is low the maximum number of offspring for high-fitness individuals is low, so low-fitness individuals are almost as capable of reproducing as individual with high-fitness. When the selection pressure is high, low-fitness individuals are less likely to reproduce and have fewer descendants. Thus, a careful balance should be maintained. If the selection pressure is too high, the genetic algorithm will converge quickly to a small number of high-fitness individuals. This phenomenon is called **premature convergence**, and we will see that it is one of the main obstacles to the success of a genetic algorithm. Premature convergence means that we may have given up too quickly on individuals with low-fitness who still could have made positive contributions to the gene pool. On the other hand, if the selection pressure is too low, we might encounter **slow convergence**, or the algorithm may not converge to a solution at all.

When a genetic algorithm starts executing, we expect to find a small number of successful individuals in a population of average and below-average individuals. It is important not to lose the better solutions at this stage. Later on, it is likely that there will still be a variance in the population, but the average fitness of the population may be close to the maximal fitness. Nonetheless, we would like the higher-fitness individuals to have a larger influence on the next generation than the average-fitness individuals.

The standard selection mechanism used most often by genetic algorithms is called **roulette wheel selection** and emulates the game of roulette, where each individual is represented by a set of consecutive slots on the roulette wheel with a size relative to its fitness, as seen in Figure 3.3. Choosing this roulette system to select parents gives every individual a probability to become a parent, which is the ratio between its fitness and the sum of all the individual fitness values.

Note that with the roulette mechanism there is a danger of early convergence. The higher the fitness value of an individual, the more often it will be selected as a parent and therefore will have more descendants. If the initial population has too many high-fitness individuals, they will be selected as parents most of the time, and we may lose solutions.

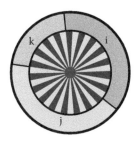

FIGURE 3.3 Roulette wheel selection. Here individual j has the highest fitness, while individual k has the lowest fitness. Therefore, when the wheel spins, the probability that j is selected as a parent is the highest, k has the lowest chance to become a parent, and i has an intermediate probability.

Several alternative selection mechanisms may be more suitable in certain cases, including the following:

- **Rank selection**: In this technique individuals are sorted by their fitness and chosen as parents based on their ranking rather than directly by their fitness. This method helps to avoid early convergence since the selection pressure is low, and low-fitness individuals may be selected quite often as parents. The disadvantage of this mechanism is that an individual might have a significantly higher fitness than the next ranked individual, yet ranking hides such differences in fitness and may cause slow convergence. Conversely, ranking works well when the fitness differences are small but believed to be significant. The roulette mechanism will select each individual with a similar probability, but the ranking mechanism can give rise to more significant differences between the selection probabilities.

- **Tournament selection**: In its most naïve form this mechanism works as follows. Two individuals are randomly chosen from the population. The one with the higher fitness is then chosen to serve as parent (the two individuals may or may not be returned to the population so they can be reselected). There are more general variants of this mechanism where the number of individuals compared (the tournament size) is larger than two. One advantage of tournament selection is that it is somewhat easier to implement than more complicated selection regiments while still allowing the adjustment of selection pressure (by choosing different tournament sizes). Another advantage is that there is no need to precompute the fitness values, which require evaluating of the entire population, since the contestants can be compared based on the evaluation function. This eliminates the need for two passes through the population to compute fitness or the cost of sorting the population (in rank selection). Eliminating the need for global population statistics makes tournament selection better suited to parallel implementations. Similarly, there are cases in which it is easier to compare solutions, as done in tournament selection, rather than to compute an evaluation function for individual solutions (e.g., if solutions represent different game playing strategies, pitting them against each other may be the best way to evaluate their quality).

- **Steady-state selection**: In this mechanism only a small fraction of the population is replaced at every generation, whereas most individuals continue to live from one generation to the next. Usually the low-fitness individuals are replaced, of course. After deciding which individuals are to be replaced, the next generation's parents have to be selected using one of the selection mechanisms so that new individuals can be created to replace the ones to be eliminated. In this method the turnover between generations is much more gradual. It is interesting to note that with this mechanism better solutions survive from generation to generation without the danger of crossover.

Each of these mechanisms has many variants that are appropriate when solving specific problems.

3.2.2 Variations on Fitness Functions

We have assumed up to now that the values of the evaluation and fitness functions are *positive* (or at least nonnegative) and that higher fitness values represent better solutions, that is, that the goal of the genetic algorithm is to *maximize* the fitness function. For many problems these assumptions are not appropriate.

Sometimes fitness values are not necessarily positive. For instance, when we search for the maximal value of the function $g(x,y) = x^2 - y^2$, for some ranges of x and y the function has only negative values. In this case, if we use the roulette mechanism the fitness values will give us incorrect roulette cell sizes.

Moreover, many problems are represented as *minimization* problems rather than maximization problems. In most of the problems we discussed, the evaluation function represented the level of success of the solution; therefore, our goal was to find a solution of the highest evaluation value. In many other problems the evaluation function represents the *cost* of a solution; therefore, we want to minimize its value.

We describe a couple of techniques that allow us to use genetic algorithms in cases where the natural fitness measure is not suitable for the selection mechanisms in their standard form. In all such cases the solution is to map the values of the evaluation function to values that are better suited to serve as fitness values.

- **Minimization problems**: The common solution is to subtract the value of the evaluation function from some constant:

$$f(x) = \begin{cases} C_{\max} - g(x) & g(x) < C_{\max} \\ 0 & \textit{otherwise} \end{cases}$$

We can select $C_{\max}$ in advance (as an input to the system), or as the largest value of g observed so far, or as the largest value of g in the current generation, or in the last k generations.

- **Negative evaluation function values**: Here the solution is to add to the evaluation function $g(x)$ a large positive value

$$f(x) = \begin{cases} g(x) + C_{\min} & g(x) + C_{\min} > 0 \\ 0 & \textit{otherwise} \end{cases}$$

As in the previous case, $C_{\min}$ may be selected in advance (as an input to the system), as the absolute value of the worst g observed so far, or observed in the current generation, or observed in the last k generations.

- **Dynamic range**: In the implementation we presented so far we used the same mapping of evaluation function to fitness function for the entire duration of the algorithm. This may not always be suitable. Consider the following example. Assume that the evaluation values are distributed between 1 and 10, and thus the best solution (10) will have a huge advantage over the worst (1) when using roulette wheel selection. Now assume that all solutions were improved by a constant value, and the range is now between 1001 and 1010. Now, the best solution (1010) will have only marginal advantage over the worst (1001). To avoid this problem and to keep supplying the impetus to continue to push the better solutions further, we transform the evaluation values into the range between the best and the worst values by the following normalization:

$$f(x) = \frac{\varepsilon + g(x) - C_{\min}}{C_{\max} - C_{\min}}$$

where $C_{\min}$ and $C_{\max}$ are the current minimal and maximal values in the population, respectively; the ε is added to make sure that even the solution with the minimal value will still have a nonzero probability to be selected.

These techniques help with some of the more common situations. More sophisticated mappings are also possible, but they need to be evaluated to see how well they really improve the behavior of the algorithm (e.g., how well they help avoid early convergence and slow termination).

For the most part we assume that individual fitness (i.e., the value of the evaluation function) is computed independently for each individual. There are, however, situations in which it is either necessary or profitable to compute the fitness of individuals based on how they interact with other individuals. An interesting technique is based on allowing individuals from two populations to compete with one another. For example, if the goal is to evolve a mechanism that handles arbitrary data in an appropriate way, it might be useful to evolve two populations: one consisting of candidate mechanisms, and another consisting of datasets that attempt to cause the mechanisms to fail. These two **co-evolving** populations will then find themselves in an evolutionary **arms race**—of the sort that can be found between predator and prey in nature—that may improve the solutions the genetic algorithm will manage to find. More generally, the fitness function does not have to be a fixed and simple function specified in advance but may in fact be a complex algorithmic function whose result changes depends on the population of solutions or on other factors. See the exercises at the end of this chapter for examples. We return to this topic in the discussion of Artificial Life in Chapter 6.

3.2.3 Genetic Operators and the Representation of Solutions

The two genetic operators we have used—mutation and crossover—expand the set of solutions tested by the algorithm. Mutations change the chromosomes in a random fashion, and their influence on the fitness of individuals is determined in the following generations. Crossover is more significant, as it allows the genetic algorithm to combine elements of already reasonably good solutions to create better solutions. Other genetic operators for improving the performance of genetic algorithms for certain problems have been suggested over the years.

It is important to realize that one of the crucial factors determining how well a genetic algorithm performs is the representation of the solutions as chromosomes—that is, the way the evaluation function interprets the chromosome—and how well this representation is compatible with the genetic operators. As the representation of solutions is such a central issue, it is sometimes called the **representation problem**.

Consider the example of designing a genetic algorithm for finding values for $x \in [31,\ldots,62]$ and $y \in [0,\ldots,31]$, which maximize the value of the function $g(x) = x^2 - y^2$. In this case it is clear that, given a possible solution (x,y), any change that will increase x will yield a better solution, as will any change that will decrease y. Thus, the changes in x and y should be independent and the representation should provide this independence. The "natural" way to achieve this is to represent the variables as a chromosome of length 10, where the first five bits represent the first gene x (as $x - 31$) and the last five bits represent y (the second "gene"). How will this representation behave when the genetic operators are applied? Mutations (which change a single bit) will obviously operate on x and y independently. What about crossover? Look at the example in Figure 3.4: assume crossover happens at the mark in Figure 3.4(a), and the result of the crossover is shown at Figure 3.4(b). In both possible offspring x is changed, while the y's remain as they were in the parents' generation. Note that for all possible crossover points x and y are never both affected by a single crossover.

If we had chosen a different representation for x and y we could have created a situation in which a crossover could affect the values of

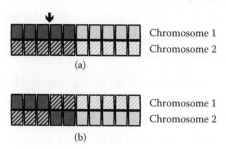

FIGURE 3.4   Effect of crossing over on two chromosomes.

FIGURE 3.5   Interlaced representation of two variables.

both variables simultaneously. Consider, for example, a representation of solutions in which the bits representing the two variables are interlaced (Figure 3.5). What will happen when we cross over two solutions under this representation? By changing the representation we created a situation in which the crossover operator changes the values of both genes. Therefore, we have increased the rate of generating new solutions, but we may pay for this by damaging parts of the solution (e.g., $y = 0$) that may have been found by the algorithm earlier. In fact, note that offspring may not carry any significant property of either of their parents. Is such a payoff worthwhile? Usually we can assume that the answer to this question will be negative (this is clearly the case for the function g) and will result in lengthening the search process, often significantly.

Another example of the difficulty of choosing a representation is manifested by the genetic algorithm solution to the following **clustering** problem. Given a set of N elements (vectors in R^n), arrange them in K sets (where $K \ll N$), and define one of the elements in each set as its *center*, such that over the K sets the sum of the distances between the elements and the centers of the clusters they belong to is minimized. There are two natural ways to represent the solutions:

1. Every solution is represented as a vector of length N, where every element denotes to which of the sets $1,\dots,K$ the corresponding element belongs. The center of each set can be calculated by looking for the element whose sum of distances to all other elements in its set is minimal. Note that in this representation the membership of each element is explicitly represented but that the identity of the centers is implicit and must be calculated in each generation.

2. Alternatively, each solution can be represented by a vector of length K that holds the identity of the center of each set. An element belongs to set k if its distance from the center of the set is minimal compared with its distances to all other centers. Here, the identity of the centers is explicit, but the membership of the elements has to be calculated in each generation.

In this case it is harder to determine which representation leads to a more effective evolutionary search without implementing both representations. It turns out the second representation is significantly more efficient.

Various genetic operators were proposed to improve the performance of genetic algorithms for specific problems. Many of these were generalizations of the crossover operator. An example is increasing the number of break points along the chromosome: for example, taking the first segment of the chromosome from one parent, then the second from the other parent, and then the last part again from the first parent. In the limit, each bit can be taken independently from either parent. Another suggestion was to use the information from more than two parents in generating each offspring. In the exercises at the end of the chapter we describe other representation methods and the corresponding genetic operators.

So far we have presented the classical genetic algorithm, but the generic evolutionary algorithm can be implemented using different representations and different operators. For instance, the genetic information may be represented as a vector of real numbers rather than bits; mutations may be implemented by adding a random number (chosen from some distribution, say, the normal distribution); the rate of mutations, P_m, may change during the evolutionary process; and other genetic operators in addition to crossover may be used. Among these operators we will mention the **inversion** operator, which selects two points on the chromosome and inverts the bit sequence between them (e.g., 11101 becomes 10111 when the underlined sequence is inverted). This operator is inspired by chromosomal inversions that happen in nature. For this operator to be useful, the "meaning" of the bits has to be independent of their location on the chromosome; that is, every bit has to have an identity. Given such an identity, inversion will change the order of the "genes" on the chromosome causing a change in the distance between genes, which might impact crossovers significantly. For example, if the problem has 10 parameters a,b,...,j, each of which can be 0 or 1, we can think of each bit in the chromosome as the "gene" for one of the parameters. It might be that ordering the genes on the chromosome abcfedghij will lead to better performance than abcdefghij, since abcf might tend to vary together because values for these genes represent a component of good solutions in the search space. In other words, different regions of the chromosome will become building blocks for generating solutions. Inversion is particularly useful when the representation of individuals is richer than a simple string of bits.

It is of course possible to come up with other genetic operators, either by imitating other biological processes or based on theoretical considerations. The appropriateness of the operators to particular problems should then be evaluated by gauging their success in improving the solutions found by the evolutionary search process and by examining the behavior of the algorithm (e.g., its speed, the number of generations it takes it to converge, the risk of converging to local maxima).

We can see from the previous examples that it can be very difficult to guess in advance how well a particular crossover method will work for a given evaluation function, and the decision is based to a large extent on intuition, understanding of the problem, and trial and error. Note that in the true biological context the "evaluation function" (i.e., the success and adaptation of an organism) is not a simple mathematical function, and therefore the corresponding problem of how nature finds good representations is even harder. For example, it would be interesting to understand how evolution determines how many genes are needed to represent a certain characteristic (e.g., one gene with a few functions as opposed to several independent genes), how close to each other in the genome should related genes be located, and how evolution "decides" whether to preserve one copy of a gene or a few copies in different locations to protect against a change in one of the copies (on this issue see also the discussion of robustness of gene networks in Chapter 6).

3.3 EXAMPLE APPLICATIONS

This section presents a few domains in which genetic algorithms have been used successfully.

3.3.1 Scheduling

Scheduling problems, from arranging a weekly course schedule for college students to designing the production floor of a car factory, are multiobjective optimization problems with multiple inputs, multiple constraints, and limited resources. Such problems are very difficult to solve using standard optimization methods. Genetic algorithms can address, in a single framework, various subtypes of optimization problems. The algorithm maintains a population of alternative schedules, and mixing and matching improves their overall performance. Thus, genetic algorithms have become a popular method to address such problems (Bagchi, 1999). More generally, an important question about genetic algorithms and similar techniques is how best to approach problems that require optimizing several objectives simultaneously (Coello, 2000). A naïve way of working with multiple objectives, and hence multiple fitness functions, is to aggregate them into one multi-objective function using a weighted sum. Note that for this to work all functions should be approximately in the same numerical range, or they must be normalized appropriately. In another interesting technique, called the **vector evaluated genetic algorithm** (VEGA), for a population of size M, k subpopulations (where k is the number of objectives) of size M/k are evaluated. Individuals are selected in each subpopulation according to a fitness function that is based on a *single* objective. The populations are then mixed and the regular operations of mutations and crossing over are performed and finally the procedure is iterated for the next generation. Both the weighted sum and the VEGA techniques are easy to implement but are severely limited in the type of cases they handle successfully (Coello, 2000).

3.3.2 Engineering Optimization

In an application of genetic algorithms to an engineering problem (Goldberg, 1989) the goal was to optimize the structure of an oil pipe, which consisted of stretches of pipe and compressor units used to maintain pressure, so that energy needed to operate the compressors is minimized under the constraints of the minimal and maximal allowed pressure in each pipe segment.

3.3.3 Pattern Recognition and Classification

A very early system (Cavicchio, 1970) used genetic algorithms to find good feature detectors for an image classification device. The images are composed of 25 × 25 black and white pixels and are divided into various named classes. During the training stage, images belonging to known classes are presented to the device, and the states of subsets of pixels (which serve as detectors) are recorded. In the recognition phase an unknown image is presented to the device, which then ranks the classes to which the image may belong based on the responses of the feature detectors to the new image and the information learned during the training stage. The goal of the genetic algorithm is to find subsets of pixels (i.e., detectors) that can be used to improve the classification success. Clearly, the set of detectors (each being a set of pixels inspected by that detector) determines how well the system performs in the classification. Genetic algorithms, in which the population consisted of sets of detectors, were used to search for successful detector sets.

3.3.4 Designing Cellular Automata

Genetic algorithms are often combined with other models such as cellular automata and neural networks. Genetic algorithms can be used, for example, to find transition rules for cellular automata that lead to a desired behavior. In an interesting experiment (described in Mitchell, 1998) a genetic algorithm was used to find a transition rule performing the "**density classification task**" on a one-dimensional cellular automaton with two states and radius 3 ($k = 2$, $r = 3$). The goal in this task is that the automaton decides whether the initial configuration of a long array contains a majority of 1's, in which case the system should settle to a fixed-point configuration of all 1's, or not, in which case it should settle to a fixed-point configuration of all 0s. Recall from the discussion of Wolfram numbering (Section 2.4) that for the automaton under discussion there are $2^{2^{2r+1}} = 2^{128}$ different rules, a huge number making it impossible to search for appropriate rules exhaustively. The genetic algorithm operated with a population of 100 possible solutions each representing a rule and measured their fitness by checking how well they preformed the density classification task on a random set of initial configurations (in each generation a new set of initial configurations was used). Most of the time the genetic algorithm was unsuccessful in finding a good transition rule, but in 3% of the runs it was able find transition rules that significantly outperformed naïve strategies for solving the density classification task.

3.3.5 Designing Neural Networks

In another application that combined the use of genetic algorithm and another biologically inspired computation model, a genetic algorithm was used to evolve a neural network that is able to learn to recognize handwriting well. As will be explained in Chapter 4, neural networks have internal mechanisms to modify the connection between elements once the network has been laid out. However, they do not have an internal mechanism that can help in designing the network layout itself. Genetic algorithms can be used to make a selective competition between different layouts and thus evolve better networks (Miller, Todd, and Hegde, 1989).

3.3.6 Bioinformatics

In a typical situation, a researcher is faced with a set of DNA sequences that share a biological function and another set of sequences that do not. The challenge is to find a sequence motif (which we assume can be represented as regular expression) present in all (or most) of the sequences from the first set and none (or very few) from the second set. Several researchers have used evolutionary computation techniques to find regular expressions that represent shared properties of sets of related nucleotide sequences. In these systems the fitness of solutions depends on how well the regular expressions match the set of sequences and possibly also how well they fail to match a control set of irrelevant sequences. In Section 3.6, which deals with genetic programming, we discuss this computational task in more detail. Langdon and Harrison (2008) provide a clear explanation of one such system.

3.4 ANALYSIS OF THE BEHAVIOR OF GENETIC ALGORITHMS

We have to ask ourselves why genetic algorithms so often succeed in finding good solutions to difficult problems. We can reformulate this question and ask which problems can be solved efficiently by using genetic algorithms, as there are problems for which genetic algorithms fail to find good solutions. To answer these questions we have to understand how genetic algorithms behave as search strategies.

Search algorithms, including genetic algorithms, search for an optimal solution to a problem in a **search space**. The search space defines the possible values of the parameters that characterize a solution. Any point in the search space represents one solution. For example, if a solution is a 2-tuple of real numbers x and y, then the search space is a two-dimesional plane, where each point in the plane is a 2-tuple (x,y) representing a particular solution.

For every point in the search space we can associate a value that indicates how successful this point is (we called this value the *value of the evaluation function*). So, if, for example, the solution is the 2-tuple (x,y), we can associate it with the value $h = f(x,y)$. We can think of this value as another dimension or another axis depicting the height (i.e., the evaluation function's value) for each point in the domain. Plotting these values will display the values of the evaluation function for all the solutions in the search space. This graph (obviously not limited to the two-dimensional case) is called a **fitness landscape**. Figure 3.6 provides examples of fitness landscapes.

Think of a genetic algorithm's search process as a walk on the fitness landscape aiming at getting to the highest point. This exposes the greatest problem faced by any algorithm that "walks" or "climbs" on a landscape from one point to a close point—it can become "stuck" in a **local maximum**. That is, it can reach a point that is not the highest in

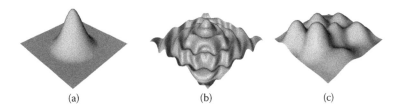

(a)　　　　　　　(b)　　　　　　　(c)

FIGURE 3.6　Different types of fitness landscapes. (a) Smooth landscape with a single maximum. (b) A structured landscape with many equivalent local maxima and a single global maximum. (c) A rugged landscape with several local maxima and a global maximum that is difficult to identify.

the entire search space; however, since it is higher than all its immediate neighbors, moving away in any direction entails going downhill. The shape of the fitness landscape in Figure 3.6(c) clearly demonstrates the problem.

To better understand the behavior of genetic algorithms, we will compare them with a more basic algorithm, called **hill climbing**, which is based on climbing a fitness landscape. This algorithm is similar to a genetic algorithm with a population of size 1, which uses only the mutation operator:

1. Start with a random chromosome x in the search space, and compute its fitness.

2. Choose the best change for x:

    a. Compute the fitness of every possible 1-bit mutation of x.

    b. Let x be the chromosome with the best fitness among all mutations.

3. Repeat Step 2 until no 1-bit mutation improves the fitness. Return x (which is the "summit" reached by the algorithm).

It is obvious that this algorithm will halt when it reaches a local maximum, even if the search space contains better solutions (solutions with higher fitness). There are many ways to address this problem. A simple solution would be to apply hill climbing from a set of different starting points (just like a genetic algorithm) rather than from a single starting point. This is called **iterated hill climbing** and is defined by adding the following to the hill-climbing procedure:

4. Return to the previous Step 1, and choose another starting point.

5. After a number of repetitions of the previous Steps 1–3, return the highest summit achieved.

The advantage of the iterated hill-climbing algorithm is clear, but note that even in this case all the climbs might end in local maxima (look again at Figure 3.6, and observe how this danger depends on the shape of the fitness landscape).

The main difference between the hill-climbing algorithms and genetic algorithms that also climb the fitness landscape toward better solutions is that genetic algorithms make use of crossovers to combine elements of two successful solutions. This helps when different segments of the chromosomes can be used as elements that can be combined in different ways to create new solutions. Thus, the search process performed by the genetic algorithm does not follow a single path, and the danger of local maxima is diminished.

However, genetic algorithms face a related problem—early convergence. The selection pressure causes the frequency of better solutions to increase with time. As a result, a situation where a single solution (or a very small set of solutions) comes to dominate the population in a relatively short time arises quite often. In fact, a population may end up containing N copies of the same solution. This might seem to be a good outcome showing that the algorithm found a successful solution. But, in fact, because we are dealing with hard problems, the probability that this is an optimal solution is small, and it is more likely that the algorithm found a local maximum. It is important to note that computationally there is no point in continuing to execute the algorithm once a single solution dominates the population. Most crossovers will be crossovers with identical copies that are of course meaningless, and changes will come mainly from mutations.

If early convergence occurs during the execution of the algorithm and we want to be able to overcome it, we first have to detect the problem. One possibility is to directly measure the variability in the current set of solutions. However, this may be time-consuming as it necessitates $O(N^2)$ comparisons between all pairs of solutions. This problem can be mitigated by randomly sampling a few individuals and comparing them.

Alternatively, we could compute the difference between the values of the evaluation function, as these values are computed in any case. If the difference between the best and worst solutions in a population is small and does not change over a few generations, it is probable that an early convergence has occurred.

It would seem that one could combat early convergence by lowering the selection pressure (i.e., the preference for better solutions); however, as we already mentioned, without preferring better solutions the algorithm cannot advance toward a solution, and maintaining the delicate balance between these two goals is almost unachievable.

Several approaches have been proposed for dealing with this problem. One could halt the algorithm every time early convergence is detected and start afresh with a new set of initial random solutions. After a predefined allocated running time, the algorithm is halted and the best solution found so far is returned.

Another approach is to raise the rate of mutations significantly when early convergence has been detected. The mutations "shake" the system up and break up the clustering of solutions. After a few generations the rate of mutation is turned back down to the original lower rate.

An interesting approach is **niching**. The population is initially divided into a few subpopulations, and the genetic operations take place only within subpopulations. After a period of such segregation, we allow cross-over among all the individuals in the population for a short period, and then the population is again divided into subpopulations and the procedure repeated. Assuming that each subpopulation will converge to a different local optimum, the periodic mixing of the subpopulation may create new solutions with the hope that some of them will combine better parts that have risen independently in different subpopulations.

A frequently arising issue when implementing a genetic algorithm is the simple linear trade-off between the size of the population and the number of generations. If you can allocate 100,000 computational steps (where a step is the evaluation of one candidate solution), you can have, for example, a population of 1000 solutions that runs for 100 generations or a population of 100 solutions that runs for 1000 steps. Clearly, we do not want to go to the limits: evaluating a population consisting of a single solution for 100,000 steps or a population of 100,000 solutions for one step makes no sense. Selecting the actual optimal trade-off point is tricky and, as many other decisions in genetic algorithms, requires trial and error. However, our experience suggests that it is often better to go with a larger population for a smaller number of generations as this decreases the probability of premature convergence.

3.4.1 Holland's Building Blocks Hypothesis

What is the source of the strength of the genetic algorithm paradigm? The main idea is that segments of the chromosome code for favorable features of the desired solution. The algorithm succeeds, from time to time, to combine such favorable segments. This is done by the crossover operation, which can take two favorable segments, each residing on a separate chromosome, and can create a single chromosome containing both segments, thereby creating a much better solution. John Holland called these segments **building blocks**. Holland expanded this notion with his notion of **schema**, a set of solutions in the search space that have a common structure. For chromosomes represented as a sequence of bits, a schema will be a template that determines only a subset of the values of the bits. The bits not determined are denoted by the character * (a *wild card* or *don't care*).

We call the bits defined by the schema (i.e., the bits that are not wild cards) **defined bits**. The number of defined bits for a schema H is called the **order of the schema** (e.g., a schema of order 1, a schema of order 2) and is denoted $o(H)$. The length of a schema, denoted by $d(H)$, is the distance from the first to the last defined bits in the schema. For example, for $H =$ **10*10*, $o(H) = 4$ and $d(H) = 5$.

Observe two individuals in a population of chromosomes consisting of four bits:

$$A = 1010$$

$$B = 1001$$

These chromosomes have a few mutual characteristics that can be denoted as schemas: 1*** (a schema representing the fact that both chromosomes have a 1 as their first bit), 10**, or *0**. When a chromosome fits a schema, we say it is an **instance** of the schema.

Obviously, a genetic algorithm acts on chromosomes and not on schemas, but we can think of the computational process as *sampling* specific instances of different schemas. This allows us to view the progress of the algorithm as a process where the values of the fitness function in conjunction with selection, and the mutation and crossover operators, divert the algorithm from schemas with low *average* fitness values toward schemas with high average fitness values.

Holland's **building blocks hypothesis** states that genetic algorithms tend to start out by identifying fitness differences defined by schemas of low order (i.e., schemas with a small number of defined bits). In successive generations the algorithm becomes more focused and succeeds in locating fitness values for schemas of higher orders (i.e., schemas with more and more defined bits), until it converges to the optimal region in the search space, which is highly enriched with high-order schemas with high values of the fitness function. The reason for this behavior is that low-order schemas will in general have more instances in the population. In other words, they will be better sampled. Low-order schemas, according to this perspective, provide coarse-grained estimates but serve as building blocks for more complicated high-order schemas (which are created mainly through crossing over), as the algorithm is drawn to regions of the search space characterized by schemas that have on average higher fitness values. According to the building blocks hypothesis this process is the source of the strength of genetic algorithms as a means for searching for solutions and optimization.

The genetic algorithm computes, indirectly, the average fitness of all schemas that have instances in the population and increases or decreases the number of instances accordingly. The concurrent evolution of a large number of schemas in a population consisting of a much smaller number of individuals is called **implicit** or **intrinsic parallelism**. This is one of the explanations for the effectiveness of genetic algorithms.

3.4.2 The Schema Theorem

In this section we prove Holland's **schema theorem**, also known as the fundamental theorem of genetic algorithms. The theorem formally characterizes the behavior of genetic algorithms.

Let $m(H,t)$ be the number of instances of schema H at time t, and let $u(H,t)$ be the average fitness of individuals who are instances of H at time t. We would like to compute $m(H,t + 1)$, the number of instances of H in the population at time $t + 1$.

Assume that the parents are chosen using the roulette mechanism. Recall that the expected number of children of individual x in the population is

$$\frac{f(x)}{\bar{f}(t)}$$

where $f(x)$ is the fitness of x, and $\bar{f}(t)$ is the average fitness of the population at time t.

Let x denote individuals in the population at time t, which are instances of the schema H. If we ignore mutations and crossovers for the time being, we get

$$E\big[m(H,t+1)\big]=\sum_{x\in H}\frac{f(x)}{\overline{f}(t)}=\frac{u(H,t)}{\overline{f}(t)}\cdot m(H,t)$$

Thus, the expected number of instances of a schema in the whole population grows by the ratio of the average fitness of individuals who are instances of the schema to the average fitness of all individuals in the population. The number of instances of a schema with a high average fitness will rise in the next generation, whereas the number of instances of a schema with a low average fitness will decrease. Note that although the algorithm does not explicitly compute the average fitness of a schema the value $u(H,t)$ appears in the previous formula.

Next we analyze how the crossover and mutation operators affect the behavior of the algorithm. It is enough to consider only the destructive actions of these operators because our goal is to derive a lower bound on the successfulness of a schema. We will compute the probability $S_c(H)$ that a schema H will still exist in the population after a crossover (at a single point)—that is, the probability that at least one of the descendants of instances of H will also be an instance of H. Let l be the length of the chromosome, $d(H)$ be the length of the schema as previously defined, and p_c be the probability of a crossover:

$$S_c(H)\geq 1-p_c\left(\frac{d(H)}{l-1}\right)$$

To understand this inequality note that there are $l-1$ possible crossover points, $d(H)$ among them are inside H, so there is a $d(H)/(l-1)$ probability that a crossover will cut and potentially destroy the schema. The probability that at least one such event occurs is $p_c(d(H)/(l-1))$. Therefore, the schema will be conserved with probability $1-p_c(d(H)/(l-1))$.

We stress again that this is a lower bound as there is a possibility that the bits coming from the other parent will recreate the schema.

Let us now consider the mutation operator. A schema H will be conserved after a mutation if and only if all its fixed bits were not mutated. A bit will not mutate with probability $1 - p_m$. As the probability of mutation in any bit is independent of all the other bits, the probability that schema H will be conserved after mutation is $(1 - p_m)^{o(H)}$.

Combining these results (i.e., multiplying the expected number of instances by the probability of the schema being conserved) produces **Holland's schema theorem**, which gives a bound on the expected number of instances of H in the population at time $t + 1$ which is

$$E\big[m(H),t+1)\big] \geq \frac{u(H,t)}{\bar{f}(t)} m(H,t) \left(1 - p_c \frac{d(H)}{l-1}\right)(1-p_m)^{o(H)}$$

3.4.3 Corollaries of the Schema Theorem

The schema theorem allows us to compute the rate of growth of the number of instances of H:

$$\frac{m(H,t+1)}{m(H,t)} \geq \frac{u(H,t)}{\bar{f}(t)} \left(1 - p_c \frac{d(H)}{l-1}\right)(1-p_m)^{o(H)}$$

Observe that if H is short and of a low order and if its fitness stays higher than the average fitness of the population, this expression is larger than 1, and the number of instances of H will grow, roughly at the rate of $u(H,t)/\bar{f}(t)$ every generation. This is an exponential rate of growth (and we are discussing only a lower bound). This is of course compatible with Holland's building block hypothesis previously presented.

However, this argument assumes that the population on which a genetic algorithm operates is a representative sample of the set of all possible chromosomes in the search space. This assumption is needed to conclude that an exponential progress rate *toward an optimal solution* will indeed happen. It is easy to see how problematic this assumption can be. For binary chromosomes of length 20 there are 2^{20} (which is more than a million) possible sequences. We usually restrict the algorithm to a much smaller population (say, 50 to 500 individuals). Obviously, the longer the chromosomes, the more critical this problem becomes.

On the positive side, recall that the schema theorem deals with only one aspect of the behavior of genetic algorithms. Given that some schemas have fitness advantage, the theorem shows that crossovers and mutations will not disrupt the growth in the number of instances of these schemas.

However, the power of genetic algorithms is also derived from the fact that crossovers and mutations can create even better new schemas from existing good schemas.

3.5 LAMARCKIAN EVOLUTION

Recall that Lamarckian inheritance is the inheritance of acquired characteristics. Neo-Darwinism denies the possibility of inheriting acquired characteristics, based on our knowledge of biological inheritance mechanisms, but this need not deter us from using this mechanism as a computational tool. Obviously, we first must consider how to incorporate the inheritance of acquired characteristics into the evolutionary computational model and second evaluate whether this mechanism improves the behavior of the algorithm.

In the standard genetic algorithms model every individual in the population is represented as a chromosome, and a fitness value is computed for every chromosome using the evaluation function. This means that usually we compute the fitness directly from the genetic information (i.e., the genotype), and we do not have a notion of an explicit phenotype. For example, we interpret the bits in the chromosome as binary numbers that directly represent the numbers we are trying to optimize. Recall that in nature the phenotype is generated in a complex way involving the genotype as well as the influence of the environment. This process can involve various forms of interactions between genes (recall the discussion of regulation in Chapter 1), the influence of the environment on which genes become active and to what extent, environmental influence independent of gene action, and more. To consider the inheritance of acquired characteristics it is helpful to distinguish between the phenotype and the genotype, since the inheritance of acquired characteristics amounts to having changes that occur in a phenotype reflected in the genotype that is passed to the next generation.

An example of a situation in which the distinction between genotype and phenotype is inviting is the design of learning systems, like the neural networks that were mentioned in Section 3.3.5 and will be discussed in detail in Chapter 4. The structure of the network (i.e., the number of elements and the way they are connected) is predesigned, but the strength of the connections (called weights) is adjusted in the learning phase of the network. If we use genetic algorithms to design such systems, we can represent the structure of the system as the genotype and can add a learning stage that can improve the performance of each network by changing the weights of the connections between the elements, and generates the mature phenotype. The fitness of the individuals in the population (i.e., the different networks) will be computed only after the learning stage. In the basic structure of genetic algorithms discussed thus far the fitness values are calculated for solutions immediately after they are formed, whereas here we calculate the fitness values after the solutions have been optimized. This is somewhat similar to Darwinian evolution where the phenotype of individuals, and hence their fitness, is affected by their life experience.

Recall that Weismann noted that it is unclear how a change caused by the environment can be coded as a genetic change. One can consider various systems where this problem will manifest itself (in Chapter 6, for example, we will see systems whose behavior is a complex result of the behavior of their components, making it difficult to find a change to the components that would lead to a phenotypically specified target behavior). However, in some computational cases this problem does not arise. For example, in neural networks all that is needed to implement Lamarckian evolution is to use the updated weights *after the learning stage* ("phenotypes") as the parent genotypes instead of the weights prior to the learning stage.

Now that we see how to incorporate the inheritance of acquired characteristics into the genetic algorithms model, we can address the second question and see what computational benefits may arise from this modification. The two outcomes we may hope for are better behavior of the algorithm (e.g., faster convergence to a solution, avoiding local maxima) and finding better solutions. In some cases that researchers studied, both these goals were achieved, but as usual there is no way to predict a priori if inheriting acquired characteristics will give rise to a better or worse system. Nonetheless, one aspect that was studied is worth mentioning.

When we allow learning to impact not only the fitness of solutions but also the genotypes of the individuals passed to the next generation, the role of the *environment* in which the solutions "learn" becomes important. Returning to biological systems, the inheritance of acquired characteristics allows directed changes caused by the environment, which are presumably adaptive, to be passed on directly to the next generation. This is in contrast to the natural selection scenario in which mutations are random and are selected when they happen to be appropriate for the environment. In the previously given neural network example, the inheritance of acquired characteristics means that a change in weights caused by learning in the current generation is passed on to the next generation. As the learned weights were adjusted to the challenges of the parents' generation, the offspring will have an advantage over randomly generated individuals in learning the *same tasks*. However, this advantage comes with a cost: researchers who analyzed the inheritance of acquired characteristics in the evolution of neural networks discovered that, when the environment changed rapidly from one generation to the next, neo-Darwinian inheritance worked better than the inheritance of acquired characteristics.

3.6 GENETIC PROGRAMMING

We end this chapter by presenting another evolutionary computational model that uses a different representation of solutions rather than the sequence of bits commonly used by genetic algorithms. As we have mentioned several times, it is sometimes easier and more natural to use representations other than binary chromosomes.

In **genetic programming** (Koza, 1992; Poli, Langdon, and McPhee, 2008), solutions are represented directly as computer programs that implement the different solutions to a problem. The fitness of each solution is determined by executing the program, usually on a number of different inputs, and by analyzing its success.

Despite the name genetic programming and its link to computer programs, genetic programming does not really deal with objects that are computer programs and with operators that change lines of code. In practice, the programs are represented as expression trees, and the genetic operators operate on trees.

Note that, although genetic programming has become a subfield in itself, the representation of the genetic information does not significantly affect the structure of the genetic algorithm itself, and it is always similar to standard genetic algorithms as previously described. The elements affected by the representation of solutions are those dealing directly with the genetic information, that is, the genetic operators like mutations and crossovers.

As a simple example of genetic programming, consider the problem of finding a simple mathematical function (or arithmetical expression) over x in [0..31] that most closely matches the points (0,0), (10,210), (31,0).

We will try to find a function of x composed of the four basic arithmetic operators (+, −, *, /) and real numbers in the range [0..31]. We will define the result of division by 0 as 0 (rather than an error), so we do not have to handle separately situations in which division by zero occurs during the evaluation of an expression.

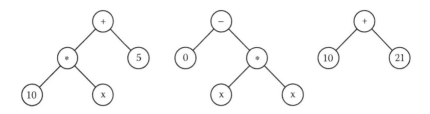

FIGURE 3.7 Trees corresponding to the arithmetic expressions $10x + 5$, $0 - x^2$, $10 + 21$.

- **Solution representation**: Each solution will be represented as a combination of the elements specified above as expression trees, such as the trees shown in Figure 3.7.

- **Initial population**: As with other evolutionary algorithms, the computation begins with a random population of candidate solutions. To generate this population we need to generate a set of random expression trees. The maximal depth of the trees is usually specified in advance and guides the process. In the **full method**, randomly chosen operator nodes are added to the tree successively until the maximum depth is reached, and beyond that only nonoperator nodes are added (in the current case these would be the numbers between 0 and 31 and the variable x). In the **grow method** both operator and nonoperator nodes are added until the maximum depth is reached (from that point on until the tree is completed, only nonoperator nodes are added). A commonly used technique called **ramped half-and-half** combines these two techniques by generating half the population using *full* and the other using *grow*. To ensure a variety of trees, various tree depths are used.

- **Evaluation function**: Every function f that is a candidate solution is evaluated by computing the values going over the list of points the function is expected to match and summing the distances between the values generated by the function and the desired results. The smaller this sum of distances, the more successful is the solution f. Note that, in contrast to most other examples in this chapter, we deal here with a minimization problem rather than a maximization problem.

- **Crossover**: Once the two parents are selected, we randomly select a node in the expression tree for each parent and exchange the subtrees of each node (Figure 3.8).

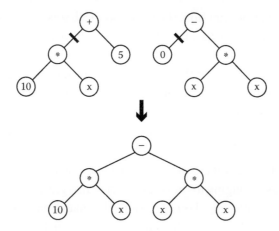

FIGURE 3.8 The crossover operation on trees. In this case $10x + 5$ and $0 - x^2$ are crossed to yield $10x - x^2$.

- **Mutation**: This is even simpler than crossover. Select a random node in the expression tree, and replace its subtree with a randomly generated subtree using the same distribution used to create the initial population. A second type of mutation, called *node replacement mutations* (or *point mutations*), is similar to single-bit mutations in genetic algorithms and consists of a random change to a single node. In mutations of this kind a function can be replaced only by a function with the same number of arguments, and leaf nodes can be replaced only by leaf nodes.

As usual in evolutionary computation, this process continues until an individual with sufficient fitness is found. In this case, for instance, we may decide to continue the processing until an individual with fitness error of at most 0.5 is found.

To summarize, the following components need to be decided in order to attack a computational problem using genetic programming:

1. The set of constants and variables that may appear in expressions—in the example we used the values [0..31] and the variable x. This set is called the **set of terminals** and is denoted by T.

2. The set of functions that are composed to create solutions (we used the four arithmetic operators). This set is called the **set of functions** and is denoted by F.

3. The evaluation function, the selection mechanism, and when will the system halt (as in standard genetic algorithms).

As a second example of the use of genetic programming, consider the following bioinformatics application. A set of DNA sequences that presumably have something in common (e.g., they may all come from regions in the DNA affected by a specific cellular mechanism) is given, and the goal is to find a regular expression (RE) that captures their similarity. To reduce the number of irrelevant answers, a control set of sequences that should not be identified as belonging to the original set is also given. Genetic programming is used to evolve regular expressions that match as many sequences in the original set and as few of the sequences in the control set.

For the sake of this example, we allow regular expressions consisting only of the four nucleotides (this is the terminal set) and the concatenation (·), alternation (|), and Kleene star (*) operators, without parentheses. The order of precedence of these operators is, from highest to lowest, Kleene star, concatenation, and alternation. We often omit the concatenation operator and write $r_1 r_2$ instead of $r_1 \cdot r_2$, thus at is the same expression as $a \cdot t$. Examples of regular expressions from this language are $a^* t$, which accepts any string of a's (including the empty string) followed by a t, and $a^* t | t^* a$, which accepts strings that belong either the previously described set of strings or strings of t's (including the empty string) followed by an a. Because of the lack of parentheses, the expression $(a \cdot t)^*$, which accepts any string consisting of repetitions of at, is not included in the specified regular expression language.

As before, solutions are represented as expression trees, crossing over is done by subtree crossing over, and mutations are done by replacing nodes with random nodes or subtrees. Clearly, a reasonable threshold has to be set for the maximal size of the tree to avoid solutions where many sequences are explicitly stored in the tree. To evaluate the quality of the different solutions, each regular expression is matched against all the sequences in each set. The quality of the solution depends on the number of true positives and true negatives as well as on the number of false positives (i.e., sequences that match the suggested regular expression but belong to the control set) and false negatives (i.e., sequences with the biological function that do not match the regular expression). To factor these four attributes in a way that takes into account the possibility that the two sets may be different in size, the **Matthews correlation coefficient** (MCC) can be used.

$$MCC = \frac{tp \cdot tn - fp \cdot fn}{\sqrt{(tp + fp)(tn + fn)(tp + fn)(tn + fp)}}$$

where *tp* and *tn* are the number of true positives and true negatives, respectively, and *fp* and *fn* are the number of false positives and false negatives.

The MCC is a value between –1 and +1, with +1 being a perfect prediction and –1 being an inverse prediction.

Note that crossing over may produce expression trees for which there is no corresponding expression in the language we specified. A crossover of g^* and $a|t$ that replaces the g node of the star expression with the entire tree of the second expression results in the tree shown in Figure 3.9.

The corresponding expression in the language would presumably be $(a|t)^*$, which is disallowed by the syntax of the previously defined language. There are various solutions to situations such as this. In some cases the problem emerges because the language is too restricted. A possible solution in this case would be to use the full-standard definition of regular expressions, assuming that in the full language such problems do not exist (or are very rare). However, there may be cases in which such a change to the language is impossible. For example, we might be using an external regular expression engine whose syntax we have to match. Alternatively, the constraints of the language may not be artificial but rather be necessary for producing expressions that make sense; for example, the language may include *if* statements in which the condition expression has to be a Boolean expression. In cases such as these we can ensure valid trees when producing the initial population and during mutation by following the syntax of the language. When doing crossover we may allow crossing over only between nodes having the same type (e.g., only a Boolean expression node can replace a Boolean expression node), or we can identify invalid

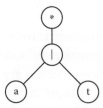

FIGURE 3.9 A tree depicting an expression disallowed by the syntax of the language.

offspring after crossover and redo the crossing over again to give the parents a chance to produce valid offspring.

This discussion should make it clear why genetic programming is formulated in terms of expression trees rather than textual source code. When textual representations are needed, parsing or unparsing to and from the tree representation may be necessary. Classically, genetic programming is formulated in terms of S-expressions, which underlie the syntax of the programming languages Lisp and Scheme. S-expressions use a prefix notation and explicitly encode the tree structure; for example, the S-expression equivalent to the arithmetical expression (2 + 3)*4 is (* (+ 2 3) 4). Since Lisp code is written using this tree notation, it is a natural candidate for genetic programming.

3.7 A SECOND LOOK AT THE EVOLUTIONARY PROCESS

In this chapter we presented the standard genetic algorithms model, which is widely used and is a basis for understanding other evolutionary algorithms. In the previous two sections we presented a few examples of using other evolutionary models for computational purposes and saw how evolutionary computation can use mechanisms such as the inheritance of acquired characteristics regardless of whether they operate in nature and to what effect. Not only are these models useful in practice, but they also provide useful intuition about cases where it is difficult to observe biological processes and thereby can help biological research by raising new questions and hypotheses (see the discussion of artificial life in Chapter 6) . In this way biology and computer science influence and enrich one another.

For this and other reasons it is useful to expand here on our knowledge of evolutionary processes. This allows us to see the genetic algorithms model we discussed in a wider perspective.

3.7.1 Mechanisms for the Generation and Inheritance of Variations

The neo-Darwinian theory presented at the beginning of this chapter stipulates that two mechanisms generate evolutionary changes: mutations and sexual reproduction. Both these processes are inherently nondirected (i.e., cannot create targeted genetic changes in response to environmental challenges faced by the organism). Mutations are random and independent of the environment. Sexual reproduction creates new combinations of existing genes and might give an advantage to certain characteristics in the population (e.g., because their carriers are more attractive sexually and therefore will have more mates and offspring) but does not create new genes, only gene combinations. The roles of the organism and the environment are very clear: the organism presents a possibly new genetic constitution, and the environment selects among individuals. This paradigm is encapsulated by the slogan, "The organism proposes and the environment disposes."

One could deduce from this that the genes uniquely determine the characteristics of the organism, and therefore the environment really selects between genes. But, as we all know, biological systems are much more complex; even identical twins—who share the same set of genes—are not completely identical. The development of an organism is a long and complex process, and the environment plays a role in most of its stages. Rather than considering development as an *inputless* computer program, we must at least realize that input from the environment can affect the development process significantly. For example, some critical developmental stages necessitate environmental stimuli or specific environmental conditions (e.g., sunlight, gravitational pull). Changes in the environment caused by the organism can affect it later on, so the relationship is interactive. No less important is the developmental flexibility exhibited by most organisms that allows them to thrive despite certain environmental conditions or constraints. The observations made by Slijper in the 1940s of a goat born without its front legs, a situation that would usually spell imminent death, are a particularly instructive example. Amazingly, the goat learned to walk on its hindquarters. After it got killed in a road accident, an autopsy was performed, and substantial changes were found in the goat's skeletal structure, its musculature, and joints: all its systems had adapted to walking upright, although it is clear that no prior genetic information in the goat's genotype was waiting to take over in such a situation, since it is unlikely that evolution had to deal with this predicament frequently enough and for enough generations.

Such examples demonstrate that the role of the environment is not limited to selecting individuals and that it has a significant impact on the development of organisms. The capability of organisms to adapt to changing conditions is called **developmental plasticity**, and, as the goat story suggests, can have a major impact on the fitness of individuals (and hence on their evolutionary success). Developmental plasticity is a mechanism for creating *variations*. However, it is important to note that in the absence of other mechanisms these changes are not inherited, and therefore developmental plasticity is not a mechanism for creating *heritable variability*.

Having compared the genetic information with a computer program, it is important to keep in mind that biological control is rather more complicated than the analogy with a sequential computer program would have us believe. Instead of a master control program, each cell and system operates semi-independently of others, and indeed so do the proteins and gene networks within each cell. The control is distributed and reactive to stimuli on a variety of levels of organization, and the resulting behavior is a complicated outcome of many localized interactions. This fact is relevant for understanding many of the phenomena we discussed, from developmental plasticity to the difficulties in the notion of inheriting acquired characteristics. In Chapter 2 we saw how simple local rules can lead to complicated and coordinated behavior. We return to this topic again in Chapter 6.

The genetic algorithms model discussed in this chapter is based on our understanding of the evolutionary process founded on genetic inheritance. Other mechanisms allowing information passing between generations are used in many situations. A fundamental example is the capability of cells in multicellular organisms to differentiate into different cell types such as nerve cells and blood cells. Recall that, since all cells carry the same genetic information, the differences between the cell types occurs because a different set of genes is expressed in each type. In general, the decision of which genes will be expressed in each cell in any given time is not based on changes in the genetic information itself (the DNA sequence) and is therefore called **epigenetic**. The epigenetic variations that create the phenotypic differences between the cells are dependent on developmental conditions for each family of cells. The changes can remain stable for long periods of time in the life of the cell and can be passed on via various processes collectively called **epigenetic inheritance** to the next generation of cells: for example, liver cells will create new liver cells. Another way for one generation to impact the next generation is by creating environmental changes. For instance, the dams a beaver builds impact the environment of its descendants and thereby the selection pressure and developmental cues they will have to deal with. It is interesting to observe that such environmental changes (called **niche construction**) impact not only the descendants of a particular individual but also the offspring of other individuals and even of other species. In this way the shared environment allows different species to mutually affect each other over many generations and for sustained periods of time.

Another route for passing information between generations is **social learning** and imitation. Social animals, including humans, imitate and learn from one another, which allows variations to spread within and between generations. As a final example, note that another way of passing information between generations is inheriting immunity to pathogens from the mother. In many species of mammals the mother passes antigens via the placenta and milk. Passing on this information (which has mainly been acquired by the mother during her life as a reaction to infections and is thus similar to passing acquired characteristics), allows the descendants to gain protection from pathogens in the environment without the dangers associated with infection.

Eva Jablonka and **Marion Lamb** proposed that the existence of such inheritance processes in addition to genetic inheritance suggests that in some cases heritable variations of these kinds are more advantageous than genetic variations (Jablonka and Lamb, 2005); we should ask ourselves how we can adopt similar processes for evolutionary computation.

3.7.2 Selection

Selection is the mechanism that turns random variation into cumulative evolutionary change. Throughout this chapter we assumed that selection is the main mechanism for evolutionary change that leads to individuals adapted to their environment. The question as to whether this really is the main mechanism leading to evolutionary change in nature and how important other processes, such as the previously described variational mechanisms, are to the direction of evolution is a major area of debate between researchers. A related debate is on the importance of processes in which members of a random subset of a population, not determined by fitness, become parents to the next generation. We can consider such processes as sampling processes, since only a random sample of the population becomes the genetic source of the next generation. A simple example of this occurs when the environment changes significantly enough so that no individual is better adapted to the new conditions than other individuals and most of the population perishes. The remaining population is composed of the "luckier" individuals and not necessarily those with a better genetic makeup. Another example is the situation where a small random sample of the population moves to a distant and isolated location (e.g., an island) and continues reproducing there (the technique of niching introduced above is somewhat similar to this situation). If the original and new environments are different, then the resulting differences between the two populations may be due to selection. However, even if the two habitats are identical, differences between the two populations may appear because genetic variability that exists in the island population is small relative to the genetic variability in the original population. Sampling processes such as these are called **drift**. Their role in evolution and relation to selection remain controversial. Clearly, the smaller the population, the higher the risk that random drift will affect the evolution of the population. To simplify analysis, many theoretical models of evolution assume an infinite population, but in many real-life situations population sizes are in fact small. Experience shows that population size is also an important factor in evolutionary computation.

Finally, note that the fitness landscape in nature is not fixed, in contrast to the way we portrayed it throughout this chapter. In fact, it changes due to changes in the environment, the impact of other organisms in the environment, and even due to changes in the environment that are the result of the activities of the organism (e.g., beaver dams). All these factors may be incorporated into evolutionary computational techniques or arise there because of unforeseen interactions between elements of the computational model. It is not a simple climb on the fitness slope since the slope shifts and changes under the organisms' feet!

3.8 SUMMARY

We discussed computational strategies inspired by biological evolution and in particular by natural selection. Most of the discussion in this chapter was devoted to the classic genetic algorithm developed by John Holland. We described a range of variations on the basic algorithm, which in some cases show superior performance. As we saw, the representation of solutions can have a significant impact on the success of the evolutionary search, and many problems are naturally represented in formats other than binary chromosomes.

We showed how the behavior of genetic algorithms can be analyzed formally and explained the proof of Holland's schema theorem. This discussion elaborated on Holland's building block hypothesis and the implicit parallelism that characterizes genetic algorithms.

The generic structure of evolutionary computation lends itself naturally to many variations. We discussed two major ideas—Lamarckian evolution and genetic programming—but many other models are described in the literature.

The chapter concluded with a short digression about evolution in nature, with the goal of showing the richness of the evolutionary processes in nature, large parts of which are not represented directly in the classic genetic algorithm model.

3.9 PSEUDO-CODE

```
// Generic code for implementing a simple genetic algorithm

INIT_POPULATION(pop)    // Create initial population

WHILE not END_CONDITION(pop)
 BEGIN
 // The end condition can be any of the conditions described in
 // section 3.2

      REPORT(pop)              // report properties of current population

      new_pop:=NEXT_GENERATION(pop) // create next generation
      pop:=new_pop
 END
REPORT(pop)              // report properties of final population
```

```
// Generic code for generating a new generation

NEXT_GENERATION(pop)

      new_pop:={}

      CALCULATE_FITNESS_OF_INDIVIDUALS(pop)

      WHILE not FULL(new_pop)
       BEGIN
            // Select a pair of individuals for mating
            parent1:=SELECT(pop)
            parent2:=SELECT(pop)
            offspring:=CROSSOVER(parent1,parent2) // Depends on Pc
            offspring:=MUTATE(offspring)           // Depends on Pm
            INSERT(new_pop, offspring)
       END

      RETURN new_pop
```

```
// A simple way to implement roulette wheel selection
// Method: Step through the wheel until the sum of fitness values
// encountered is greater or equal to a random stopping point. By
// summing fitness values we in effect use the fitness value as the
// size of each slot in the wheel.

// rand() returns a random number in the interval [0,1]
// Fitness is non-negative.

SELECT(pop)

      // sum the fitness values of all individuals
      total_fitness:=SUM_FITNESS(pop)
      wheel_location:=rand() * total_fitness

      index:=1  // point to first individual in pop
      current_sum:= FITNESS(pop[index]) // used to sum fitness values

      WHILE (current_sum<wheel_location) and (index<population_size)
       BEGIN
            index:=index+1
            current_sum:=current_sum+FITNESS(pop[index])
       END

      RETURN pop[index]
```

3.10 FURTHER READING

Bagchi, Tapan P. 1999. *Multiobjective Scheduling by Genetic Algorithms*. Norwell, MA: Kluwer Academic Publishers.

Cavicchio, Daniel Joseph. 1970. Adaptive search using simulated evolution. Ph.D. thesis, University of Michigan.

Coello, Carlos A. 2000. An updated survey of GA-based multiobjective optimization techniques. *ACM Computing Surveys* 32, no. 2, 109–143.

Doerr, Benjamin, Nils Hebbinghaus, and Frank Neumann. 2007. Speeding up evolutionary algorithms through asymetrical mutation operators. *Evolutionary Computation*, 15, no. 4 (December 1): 401–410.

Goldberg, David E. 1989. *Genetic Algorithms in Search, Optimization and Machine Learning*. Reading, MA: Addison-Wesley.

Holland, John H. 1975. *Adaptation in Natural and Artificial Systems: An Introductory Analysis with Applications to Biology, Control, and Artificial Intelligence*. Cambridge, MA: MIT Press.

Jablonka, Eva and Marion J. Lamb. 2005. *Evolution in Four Dimensions: Genetic, Epigenetic, Behavioral, and Symbolic Variation in the History of Life*. Cambridge, MA: MIT Press.

Koza, John R. 1992. *Genetic Programming: On the Programming of Computers by Means of Natural Selection*. Cambridge, MA: MIT Press.

Langdon, William B. and Andrew P. Harrison. 2008. Evolving regular expressions for GeneChip Probe performance prediction. In *Proceedings of the 10th International Conference on Parallel Problem Solving from Nature: PPSN X*, 1061–1070. Dortmund, Germany: Springer-Verlag.

Miller, Geoffrey F., Peter M. Todd, and Shailesh U. Hegde. 1989. Designing neural networks using genetic algorithms. In J. Schaffer (Ed.), *Proceedings of the Third International Conference on Genetic Algorithms*, 379–384. San Francisco: Morgan Kaufmann.

Mitchell, Melanie. 1998. Computation in cellular automata: A selected review. In T. Gramss, S. Bornholdt, M. Gross, M. Mitchell, and T. Pellizzari, *Nonstandard Computation*, 95–140. VCH Verlagsgesellschaft.

Mitchell, Melanie and Charles E. Taylor. 1999. Evolutionary computation: An overview. *Annual Review of Ecology and Systematics*, 30, 593–616.

Poli, Ricardo, William B. Langdon, and Nicholas F. McPhee (with contributions by John R. Koza). 2008. *A Field Guide to Genetic Programming*. Available at: http://lulu.com and freely available at http://www.gp-field-guide.org.uk.

3.11 EXERCISES

3.11.1 Evolutionary Computation

1. Discuss how a genetic algorithm should deal with the situation where, after a large number of generations, no good enough solution for a problem has been found. Try to suggest ways for a system designer to attempt to increase the probability of finding a solution when such a situation occurs.

3.11.2 Genetic Algorithms

2. Is it a good idea to choose a large p_m (close to 1)?

3. In the general algorithm we described, p_m determined whether to flip a bit. In another description of genetic algorithms, p_m determines whether to replace the bit with a new bit, and if so the new bit is created as a 1 with probability 0.5 and as a 0 with probability 0.5. Are these two algorithms fundamentally different, or are they the same up to simple adjustments of parameters?

4. Describe in detail how to find the values of $x \in [31,..,62]$, $y \in [0,..,31]$, which minimize the function $g(x,y) = x^2 + 2xy - y^2$ using a genetic algorithm. Consider how the selection method and population size may affect the rate of convergence and minimize the dangers of premature convergence.

5. An ant walks on a rectangular two-dimensional grid containing obstacles. All steps are of length 1 on the grid. Before each step the ant can change its direction to one of the four possible directions (north, south, east, or west). A path is composed of ten such steps (consisting of a turn and a move). When an ant hits an obstacle it cannot move further and remains stuck in place.

    a. Propose a way for representing possible paths as a chromosome of bits.

    b. Show how to use a genetic algorithm to find the maximal Euclidean distance from the beginning of the path (i.e., the maximal distance between the start and end points of a path) an ant may reach in ten steps.

3.11.3 Selection and Fitness

6. After an individual is chosen to serve as parent, it may either be discarded or returned to the population so that it can be selected again. How does this choice affect selection pressure (for concreteness, consider each selection mechanism separately)?

7.

    a. Express the probability $P(i)$ of choosing individual i as a parent using f_i $(i=1..n)$, for roulette wheel selection.

   b. Using (a) derive the expected number of offspring for individual *i* assuming the size of the population is *n*.

8. Show that the values of the probability function *P(i)* derived for roulette wheel selection are between 0 and 1 and that their sum is 1.

9.

   a. In Tournament selection does the selection pressure increase or decrease when the size of the set *k* is increased?

   b. If two individuals are selected to be compared and the individual with the higher fitness is chosen as parent with probability *p* while the individual with the lower fitness is chosen as parent with probability 1 – *p*, will the use of a smaller *p* increase or decrease the selection pressure?

10. We can add an **elitism** mechanism to any of the selection methods, whereby in each generation a number of individuals with high fitness are chosen and passed as is to the next generation (similar to what happens to most individuals in steady-state selection). One of the advantages this offers is that it guarantees the monotonicity of the maximal fitness in the population (as it clearly can only increase). Does this raise or lower the selection pressure?

11. Try to find biological examples analogous to each of the selection techniques presented. Try to speculate whether these techniques were developed based on biological observation or as an attempt to improve the performance of genetic algorithms.

3.11.4 Genetic Operators and the Representation of Solutions

12. A chromosome is composed of two consecutive genes *A* and *B*, each of which is represented by five bits. There are two values (alleles) of *A* and two alleles of *B* in the population. The two alleles differ in every bit. Each allele of each gene appears in 50% of the population. What is the probability that a crossover between two chromosomes will give rise to a new allele of *A* that did not exist in the original population?

13. Repeat the analysis in question 12 for chromosomes of length 100, where *A* is on the left end of the chromosome, *B* on its right and there

are the 90 bits between them all with value 0 for every individual in the population.

14. One can change the "rate" of mutations p_m during the execution of a genetic algorithm. What would be the advantage of doing so? How should one change p_m as the execution progresses?

3.11.5 Analysis of the Behavior of Genetic Algorithms

15. Given the schema $H = 1{*}{*}{*}01{*}$, compute $o(H)$ and $d(H)$. How many possible instances are there of H?

16. Compute the number of schemas for which a given chromosome of length l can be an instance.

17. Prove that not every subset of chromosomes of length l is uniquely defined by a schema.

18. Assume a genetic algorithm or hill-climbing algorithm that represents the solutions as a real number x rather than a chromosome made up of bits. How would you implement mutations in this representation? (Hint: think of the definition of the derivative.)

19. Explain why the niche method lowers the risk of early convergence.

20. The following observation lowers the bound given by the schema theorem: a crossover cannot destroy a schema if both parents are instances of the schema, regardless of the crossover point. Refine the bound we derived using this observation. Hint: derive an expression that computes the probability that at time t the second parent will belong to H, under the assumption that the first parent belongs to H. We are interested in the complement of this situation. Combine this expression with the expression dealing with crossovers in the proof we presented for the schema theorem.

21. We can deduce from the schema theorem that the success of a particular schema is independent of the success of other schemas in the population and depends only on its average fitness and the average fitness of the population as a whole. Explain this statement and why it supports the idea of implicit parallelism. Based on the discussion of the schema theorem in the chapter, qualify this conclusion.

3.11.6 Genetic Programming

22. Consider a case in which genetic programming is used for a problem of finding a mathematical function that goes through a given set of x,y points (numerical regression). Suppose additionally that the solution must be of the specific form such as $a*sin(x) + b*cos(x)$, where a and b may be any arithmetic expressions not involving x. Suggest ways to use genetic programming to find solutions that have the desired form.

3.11.7 Programming Exercises

23. Given 10 cards numbered 1 through 10, divide them into two piles, such that the sum of the card values in the first pile will be as close as possible to 36 and the product of the card values in the second pile will be as close as possible to 360.

    a. Suggest a way to solve this problem using a genetic algorithm, and describe all its components. It is recommended to write a whole computer program to implement the evolutionary computation.

    b. Do you think that genetic algorithms are suitable for solving this problem? Did you change your mind after you experimented with the program?

24. Given x,y coordinates of 10 cities and assuming an Euclidian distance between each pair of cities, use a genetic algorithm to find the shortest path through all of the cities (the traveling salesman problem).

25. **Sorting networks** are hardware components used to sort sequences of numbers. Their advantage is their simplicity and their high parallelism. They operate as follows. Every number in the sequence

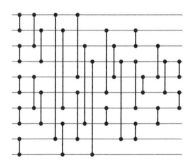

FIGURE 3.10

is put in a channel (the number of channels equals the sequence length). Every two channels are connected with a **comparator** element, a component that compares two numbers and exchanges them as needed so that the number in the lower-indexed channel is the smaller one. The numbers in the sequence move along the channels and are exchanged if need be. The goal is that at the end of execution the channels will contain a sorted sequence. The network in Figure 3.10 sorts 10 elements (the horizontal lines are the channels, and the vertical lines are the comparators). The depth of the network is defined by the number of nonparallel exchanges, so in this example the network contains 29 comparators and has a depth of 16. A good network will use a small number of comparators and will have a low depth. In this exercise we will attempt to find a network that will correctly sort any possible input and will have a small number of comparators using a genetic algorithm (you can ignore the low depth requirement).

a. Design and implement a genetic algorithm for designing a sorting network of sequences of numbers of length 12. Analyze the behavior of the algorithm in relation to the following points:

   i. How did you deal with the huge number of possible input sequences? How did you determine when to stop searching for a solution?

   ii. How did you represent the solution? What were the genetic operators, and the evaluation function, and how did these decisions impact the behavior of the algorithm?

   iii. How would you modify the representation if you were trying to also minimize the depth of the network?

   iv. What did you do to avoid local maxima?

b. To escape from local optimums, we can use the following technique: in addition to the population networks, we will run in parallel an evolutionary process for a population of test sequences that the networks have to sort correctly (the more sequences are sorted correctly, the higher the fitness of the sorting network). The "goal" of the sequences is to hinder the performance of the sorting network, so a fitness of a sequence will be higher if networks fail in sorting it. Implement this idea, and test its impact on the performance of the sorting networks. How would you describe this change in terms of what it does to the fitness landscape of sorting networks?

26. Given 100 points on the plane represented as 2-tuples (x,y), use a genetic algorithm to sort them into five clusters. Use the two representations described in Section 3.2.3 and compare the behavior of the algorithm and the quality of the results.

27. Given an undirected connected graph $G = (V,E)$, compute a cycle (a path that starts and ends on the same vertex) such that every edge is used exactly once (such a cycle is called an **Eulerian cycle** if such a cycle exists). Let m be the number of edges. We want to solve this problem using two evolutionary algorithms, with no crossover operations—that is, only the mutation operator is involved. A solution is represented as a permutation of the edges of G (and not as a binary chromosome). The fitness is defined to be the length of the longest prefix of the solution that is connected (adjacent edges that share a vertex). If the fitness equals the number of edges then the solution represents an Eulerian cycle. It was claimed that an asymmetric mutation operator may be more effective than a symmetric one (Doerr et al., 2007). In this exercise we will investigage this claim.

Solution 1—Asymmetric mutation operator:

    i. Choose a random permutation of edges π as the initial solution.

    ii. Choose $1 \le i \le m$ uniformly at random. Let π' be the result of moving the element at position i of π to position 1 and shifting the elements between position 1 and position i one position to the right. For example, jump(5) applied to (6,4,3,1,7,2,5) produces (7,6,4,3,1,2,5).

    iii. Replace π by π' if π' has higher fitness.

    iv. Repeat Steps 2 and 3 forever.

Solution 2—Symmetric mutation operator:

    i. Choose a random permutation of edges π as the initial solution.

ii. Choose $1 \le i,j \le m$ uniformly at random. Let π' be the result of jump (i,j) applied to π, where the element at position i is moved to position j while the other elements between position i and position j are shifted in the appropriate direction. For example, jump(5,2) applied to (6,4,3,1,7,2,5) produces (6,7,4,3,1,2,5).

iii. Replace π by π' if π' has higher fitness.

iv. Repeat Steps 2 and 3 forever.

Implement both solutions and compare their performance.

28. Use a genetic algorithm to solve sudoku puzzles, and analyze the effect of Lamarckian evolution of the performance of the algorithm. Given a sudoku board, choose a population of solutions randomly (the population can be of any size, but a size between 50 and 100 seems most appropriate), which will comprise the first generation. The algorithm will generate the subsequent generations. You have to decide on the following elements of the genetic algorithm:

a. Solution representation.

b. The evaluation function.

c. How to perform crossover between solutions.

d. What mutations will consist of.

e. How to select the individuals which will be passed on to the next generation.

The main objective of this exercise is to determine the influence of Lamarckian evolution on the performance of the genetic algorithm—that is, determine the effects of allowing the inheritance of acquired characteristics. To this end, we will allow every solution to optimize its fitness and the passing of the improved solutions from one generation to the next. Optimization can be implemented, for example, by the following procedure: assume you chose to represent the solution using a 9×9 matrix. A solution is modified by exchanging two cells in the matrix. Optimization can be performed by finding the pair of cells that when exchanged will give rise to the best improvement in the evaluation function. We will allow 100 optimization steps for each solution. The representation and the optimization suggested here are only examples, and you are free to choose other reasonable approaches; however, the number of optimization steps for each individual has to be to the same. Compare the following three policies:

a. The usual genetic algorithm.

b. A genetic algorithm where each solution is optimized directly, the fitness is evaluated after the optimization step, but the solutions used for generating the next generation are the original, nonoptimized solutions.

c. A complete Lamarckian algorithm, where each solution is fully optimized, fitness is evaluated after the optimization step, and the optimized solutions are used to generate the next generation.

Game boards can be found at http://www.sudokupuzz.com and many other places. The program has to deal with boards classified as easy, medium, and hard.

Compare the algorithm's performance under the different inheritance policies and different levels of difficulty of a sudoku game.

3.12 ANSWERS TO SELECTED EXERCISES

2. Usually not. If the mutation rate is too high, the properties of the good solutions will not be preserved from one generation to the next.

3. There is no significant difference. The second algorithm behaves exactly like the original for a p_m twice as small.

7. If we denote by f_i the evaluation value of i and by $\hat{f}_i$ the fitness value of i, then

$$P(i) = \frac{\hat{f}_i}{\sum_{k=1}^{n} \hat{f}_k} = \frac{\frac{f_i}{\bar{f}}}{\sum_{k=1}^{n} \frac{f_k}{\bar{f}}} = \frac{\frac{1}{\bar{f}} f_i}{\frac{1}{\bar{f}} \sum_{k=1}^{n} f_k} = \frac{f_i}{\sum_{k=1}^{n} f_k}$$

Therefore, we deduce that with this selection mechanism and definition of fitness we can use the evaluation function directly when calculating the probability of reproduction without first computing the fitness value separately. Using the probability $P(i)$ we can estimate the expected number of descendants for i: in the population of size n as

$$nP(i) = \frac{nf_i}{\sum_{k=1}^{n} f_k} = \frac{nf_i}{\overline{nf}} = \frac{f_i}{\overline{f}}$$

This is exactly the fitness value!

9.

    a. The selection pressure increases as k increases.

    b. Lowering p will decrease the selection pressure.

10. As elitism maintains the solutions with the highest fitness and reduces the number of new offspring it raises the selection pressure.

15. $d(H) = 6$, $o(H) = 3$. There are 16 ($= 2^4$) instances of the schema.

16. A chromosome of length l belongs to 2^l schemas since a matching symbol in the schema can be either the same symbol as in the chromosome or $*$.

17. There are 2^{2^l} subsets and only 3^l possible schemas.

20. A refined expression for the schema theorem:

$$E[m(H,t+1)] \geq \frac{u(H,t)}{\overline{f}(t)} m(H,t) \left[1 - p_c \frac{d(H)}{l-1} \left(1 - \frac{m(H,t)u(H,t)}{\overline{nf}(t)} \right) \right] (1 - p_m)^{o(H)}$$

where n is the size of the population.

21. The formula we derived shows that the rate of growth of the number of instances of a schema depends only on the schema's average fitness and on the average fitness of the population. So we can consider the evolutionary process as dealing with schemas rather than individuals (i.e., the fitness of schemas rather than the fitness of individuals). But

in practice the algorithm tests the fitness of individuals and only indirectly the fitness of schemas; therefore, a small number of individuals represent a larger number of possible schemas, and we have implicit parallelism. On the other hand, if the population is too small, the sampling of schemas it provides may be insufficient and hence probably biased.

23.

a. The natural way is to represent solutions as binary chromosomes of length 10. If a bit has value 1, the card belongs to the first pile and otherwise to the second pile. To compute the fitness of a solution we will construct an error function and attempt to minimize it. The function has to take into account both piles relative to their target values (the absolute value of the difference from the targets), making this a multiobjective optimization problem. The products are bigger than the sums, so the function has to account for this when combining the values (by multiplying the term for the sum by an appropriate factor).

b. The search domain is small (1024 possibilities), and the fitness landscape is not smooth and hard to climb due to the dependence between the components of the problem. Therefore, it is probable that on average a genetic algorithm will be less efficient than an ordered search or even a random search (this can be tested by increasing the rate of mutations).

24. A possible way to represent the path to the genetic algorithm is by using a chromosome containing 10 numbers that determine the order of traveling between the cities. For example: 10 1 5 3 2 6 4 7 9 8 is a path starting at city 10, then city 1, then city 5 and so forth until it ends at city 8. A mutation would be a swap of two cities, and a crossover will be cut and paste of two permutations. Note that such a crossover does not guarantee that only valid paths (i.e., a real permutation) will be produced; thus, there is a need to go over each offspring, to check its validity, and where necessary to correct it (e.g., by randomly replacing cities that appear twice by cities that did not appear at all).

CHAPTER 4

Artificial Neural Networks

U P TO NOW WE have dealt with computational models inspired by biological systems, and our discussions have been focused on their computational capabilities, that is, their abilities to solve difficult computational problems. In this chapter we will again look at a model inspired by a biological system (the brain and the nervous system), but we will focus on a specific capability of this model, namely, its learning capability.

Artificial neural networks (ANN) are a family of computational models inspired by various aspects of the nervous system and the brain. A neural net is a set of interconnected simple computational units, similar to the brain, which is composed of a large number (on the order of 10^{11}) of neurons that are interconnected. Artificial neural nets are interesting in part due to the large number of technological applications they have and in particular because neural nets are extremely useful tools for solving problems that require learning or generalizing from examples.

4.1 BIOLOGICAL BACKGROUND

The atomic units of the nervous system are **nerve cells**, also known as **neurons**. Neurons come in many different forms, but they all share some characteristics: they have components called **dendrites** that act as "antennas" and are composed of extensions that receive signals from various sources (mainly from other neurons) and another component called an **axon** that transmits output signals (Figure 4.1). A single neuron has many inputs, averaging between 1000 and 10,000 dendrites, but only a single output axon, which splits up at its end to allow it to connect to the dendrites of several other neurons. The dendrites are relatively short, but axons can be

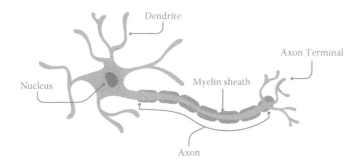

FIGURE 4.1   A schematic description of the structure of nerve cell. The dendrites are the inputs of the cell, and the long axon is the output.

very long. For instance, in humans, the axon in neurons connecting the spinal cord to the foot is about 1 meter long.

When neurons are activated, certain physical changes happen in them, and an output signal is generated. For example, photo-sensitive cells in the eye react to light by creating neural signals that are then acted upon by the nervous system and the brain. How are the signals inside a neuron and between neurons transmitted? Signals propagating along the axon are electric, whereas the signals received by the dendrites are chemical in nature. These signals are translated into electric signals by using a complex array of gates and pumps that control the ionic balance in the cell and generate electric signals. When the sum of signals in the dendrites surpasses a certain threshold, the cell body (or soma) generates an electric signal that propagates along the axon. This is called a nerve impulse. One of the reasons the signal does not decay along the axon is that the axon is sheathed by an insulating material called myelin. In patients suffering from multiple sclerosis, the myelin sheath is defective; therefore, in these patients the nervous system is severely impaired.

In contrast to this electrical conductivity, signals are passed between one neuron's axon to other neurons' dendrites using chemical molecules called neurotransmitters that cross the gap between the axon and the dendrite, called a **chemical synapse**, or just **synapse**. The sending neuron is called the presynaptic neuron, and the receiving neuron is the postsynaptic neuron. Vesicles containing neurotransmitters are at the end of the axon. The electrical impulse triggers the fusion of the vesicles to the outer membrane and changes the permeability of the vesicles, thereby causing the release of the neurotransmitters into the synapse. The neurotransmitters then defuse to the other side of the synapse, bind onto receptors on the dendrites, and cause them to activate and start the electrical signaling in the neighboring cell, or modulate it. About 100 different molecules acting as neurotransmitters have been identified in the human brain.

This complex signaling mechanism has several advantages. The translation from an electrical signal to a chemical signal and back again allows for signal amplification and prevents signal decay. Moreover, different neurotransmitters have different affinities to the postsynaptic receptors, leading, for example, to different rates of signal transmission, and as a result affect postsynaptic neurons in a differential way. Thus, the vast array of neurotransmitters allow for fine-tuned control over the function of the nervous system.

The brain size of animals varies widely—from the 0.001 gr brain of honeybees to the 10 kg brain of whales. Generally, there is a good correlation between the size (either mass or volume) of the brain and the size of the body of animals; the human brain, which typically weighs around 1.2 to 1.4 kg, stands out with a high brain-to-body ratio. However, although it is common in comparative studies to normalize the size of the brain to the size of the body, the absolute size of the brain must also play a role, as there might be a limit to how much the brain circuitry can be miniaturized. Thus, it is fascinating that honeybees, whose brain size is only about 0.001 gr (compared with a body mass of about 0.1 gr) are capable of sophisticated social behavior. Readers interested in this subject are referred to a recent review (Chittka and Niven, 2009). In general, the sophistication of the brain depends not only on its size but also on its structure and level of interconnectivity.

We already mentioned that neurons are highly connected. A neuron can receive input from tens of thousands of neurons and output signals to hundreds of other neurons. The strength of the system is due to its high connectivity. The human brain is highly connected; it consists of the order of 10^{11} neurons interconnected by the order of 10^{14} synapses, thereby allowing for complex collective computations.

The development of the nervous system can be understood as consisting of two components. First, neurons are created, and their axons grow in various directions. This determines the overall topology of the neural network. Next, the strength of connections at each synapse is refined due to the signals passing between adjacent neurons. Neurons live for a long time, and in a certain sense the brain continues to develop as long as the organism is alive. Thus, synapses are generated, strengthened, weakened, or eliminated. We do not understand all these processes in full detail, but it is known that these processes are at the core of the amazing ability of the human brain to learn and remember.

Artificial neural nets are a simplification of the biological system. We will use very simple "neurons" that can sum up their inputs and can produce a corresponding output. In these artificial networks, like real neural networks, the system gains its strength from the connections between the single cells, and the challenge is to design a system that can learn, remember, and perform complex computations.

4.1.1 Neural Networks as Computational Model

In addition to scientific curiosity, researchers try to mimic the brain when designing computational systems for many practical reasons. The brain presents many desirable properties that are hard to achieve in standard digital computational systems:

- **Fault tolerance and robustness**: Individual nerve cells can die without affecting the functionality of the system. In fact, the brain can withstand damage that is rather widespread.

- **The ability to deal with inconsistent, noisy, or unreliable data**: Our daily experience shows that the brain is capable of reaching correct decisions, at least most of the time, under conditions where the relevant data are far from being complete and entirely reliable.

- **Parallelism**: Computation in the brain happens simultaneously in different regions and is based on the local interaction between neurons connected to each other.

- **Asynchronous**: The brain does not contain a clock that synchronizes the different computational processes and nonetheless can compute effectively.

- **Learning ability**: The human brain, as well as brains of simpler organisms, can adapt the organism's behavior to changing environments. This is in stark contrast to computers, which have to be reprogrammed when computational challenges change.

4.2 LEARNING

One of the main characteristics of animals is their ability to learn. One can see this ability even in simple bacteria (which are simpler than the simplest animals) that can adapt to their environment both as individuals and as colonies (see Chapter 2). Higher organisms can perform different types of learning based on either their own experiences or by learning from teachers, such as young animals imitating their mothers, dogs being trained, or learning abstract ideas through reading as is happening between the readers of this book and its authors. Realizing the advantages of learning, scientists were motivated to design artificial systems that can learn, and this gave rise to various approaches to **machine learning.** It is customary to distinguish between two types of machine learning:

- **Supervised learning**: Animals tend to learn by example, such as when a cub learns from its mother to distinguish between dangerous and benign animals. In this case the mother is the "supervisor" that supplies the correct answers. Similarly, in a supervised learning system, the system's output is compared with answers known to be correct, and then the internal parameters are tweaked in an effort to make its output correspond better to the correct answers. The obvious goal is for the system to internalize and generalize the answers so that eventually it will be able to give correct answers to questions that are not in the training set. Supervised learning comes in two flavors: either the trainer provides the correct answer to the trainee, or the trainer merely indicates whether the solution is correct and grades it.

- **Unsupervised learning**: In addition to supervised learning, higher animals and in particular humans are capable of independent learning. In this process there is no trainer to grade the answers, and in fact there is no a priori definition of correct and incorrect answers. Still, humans learn by attempting to discover consistent patterns in phenomena they encounter. Similarly, in unsupervised learning a computational system is tasked with discovering interesting patterns in its input that have statistical significance. Discovering such patterns allows, for example, the machine to make correct decisions based on the input data and to successfully forecast future inputs.

Most of this chapter will deal with supervised learning, but we will provide one example of unsupervised learning, self-organizing maps (SOMs).

Machine learning can be implemented using standard computational techniques, but as artificial neural networks are computational models emulating natural learning systems, they are particularly well suited to such tasks. This chapter deals with such artificial neural networks and their usage as learning systems.

4.3 ARTIFICIAL NEURAL NETWORKS

4.3.1 General Structure of Artificial Neural Networks

The atomic unit of the simplest artificial neural network is an idealized neuron called the **McCulloch–Pitts neuron**, shown in Figure 4.2. The neuron can be in one of two states: (1) the **firing** state (denoted by the value 1); and (2) the **nonfiring** state (denoted by the value 0). The state of the neuron changes according to signals it receives from the neurons feeding into it: The value n_i of neuron i at time $t + 1$ is calculated by the following formula:

$$n_i(t+1) = \Theta\left(\sum_j w_{ij} n_j(t) - \theta_i\right)$$

where the function Θ is defined as follows:

$$\Theta(X) = \begin{cases} 1 & X \geq 0 \\ 0 & X < 0 \end{cases}$$

The step function Θ is called the **threshold function** or the **activation function** of the neuron.

The value θ defines the threshold required to activate the neuron, and w_{ij} are the weights that define how strongly the input from neuron j

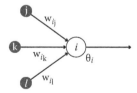

FIGURE 4.2   A simple artificial neuron. The neuron sums the inputs (each multiplied by the relevant weight). If the weighted sum is larger than the threshold θ then the neuron will have output of 1; otherwise the output will be 0.

influences neuron i. (Note that the common convention is to denote the weight leading from j to i by w_{ij}). The weighted sum of the input signals arriving at time t has to be above the threshold θ_i for neuron i to fire at time $t + 1$. The outputs of the neurons are binary values, but as these values are multiplied by the weights, which are real numbers, the inputs of the neurons are real numbers. The binary values are usually 1 and 0 (as indicated in the previous formula), although for some of the analysis that we will show it will be simpler to use the *sign* function instead of the Θ function and represent the output by 1 and –1.

A neural network is a collection of neurons that are connected to each other. Simple networks are single layered, and we will discuss them first. In these networks the neurons representing the input are connected directly to the neurons representing the output. Other networks are multilayered, which means there are one or more layers of neurons between the inputs and the outputs. Other networks have a topology that is not layered at all, for instance, a network where each neuron is connected to all other neurons. In every network every connection (often called an edge) has a weight that determines how active this connection is, and these weights w_{ij} determine how the neurons' states will change after setting the network's initial conditions.

It is easy to see that the network shown in Figure 4.3 computes the majority function: if two or more of the inputs are of value 1, the weighted sum of the inputs to the neuron will be larger than the threshold value ($\theta = 1$), and its output will be 1. For instance, if $X_1 = 1$, $X_2 = 0$, $X_3 = 1$ then the weighted sum of these values is $0.8 \times 1 + 0.7 \times 0 + 0.6 \times 1 = 1.4$, which is larger than the threshold value of 1. If only one input has the value 1, or all of the inputs are 0, the weighted average will be less than 1, and the neuron's output will be 0. In this example, the weights were chosen so the network computes the desired function, but as we will see one of the

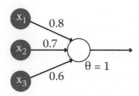

FIGURE 4.3   A very simple network comprised of one neuron that implements the majority function. If two inputs or more have the value 1 then the output will be 1; otherwise it will be 0.

main challenges in working with neural networks is to find a way for the network to self-adjust the weights so as to compute a given function.

This simple model of neurons simplifies away most of the complexity of biological neurons, but it allows us to build useful neural networks. Despite the simplicity of the single neuron, the theoretical analysis of networks of these neurons poses complicated challenges. This approach to artificial neuronal networks was suggested in 1943 by McCulloch and Pitts. They showed that it is possible to build a universal computer using such neurons, thereby demonstrating that simple units such as these are computationally strong when combined in networks.

A few of the differences between biological neurons and the abstract McCulloch–Pitts neurons are as follows:

- McCulloch–Pitts neurons follow a step function and thus have a discontinuous reaction (i.e., they don't react below the threshold and then switch to full activity). The response of real neurons is continuous although they may have rapid change of activity at some point. Some artificial neurons use continuous functions, and we will describe them and their properties when we talk about multilayer networks.

- Some real neurons address their input in a sophisticated manner rather than simple linear summation. For example, if the input from a certain source rises above a certain threshold, then this input is no longer taken into account. Examples of differential treatment can be found in the sensory system. For example, some neurons are responsible for distinguishing between sounds that reach both ears simultaneously. The neurons can distinguish between an increased volume in one ear (one input) and the same increase into both ears, even when the sum of the inputs is the same in both cases (Segev, 1998). One can model such properties of single neurons by a collection of McCulloch–Pitts neurons.

- The output of real neurons is a sequence of pulses (spikes) rather than an output of a constant level. Biological neurons can represent information not only by the level of the output but also by the rate in which the pulses are emitted. Recently artificial neural networks using such spike neurons as their basic elements have been proposed, but we will not discuss these networks here.

4.3.2 Training an Artificial Neural Network

In neural networks, learning is achieved by updating the values of the weights. As the weights determine the result of the computation, we have to set the weights so that they produce the correct output for the learning samples. In most situations this is achieved by setting initial random values for the weights and then modifying them so that the network will produce the required results. We will discuss how this process is performed in different kinds of neural networks.

In supervised learning we assume that there exists a **training set** of inputs for which the desired result of the computation is known. Although neural nets can produce numerical results (e.g., multiply two numbers), we will usually deal with networks whose computational goal is **classification**. Therefore, it is common to deal with input with a known classification into subsets, and the goal of the training phase is to teach the network to classify the input patterns by adjusting its weights. To demonstrate the relevant ideas, it often suffices to discuss a network that outputs a single bit, which classifies the data into two subsets. Obviously, by using more output neurons the data can be separated into more classes.

During the training phase, each input item from the training set is presented to the network, and the network computes the output. If the result differs from the correct answer, the weights are updated in an attempt to fix the error. This process is executed repeatedly. One pass through all the training data is called an **epoch**. As the weights are constantly being updated, there is no guarantee that input data producing a correct result will continue doing so at the end of the epoch. Similarly, there is no guarantee that the updated weights will result in a correct result when applied to an input set that had originally produced an error. Therefore, more passes through the training set may be needed. If the network produced the correct answers for all the input elements, then the training phase is completed.

We will see that there are problems and learning algorithms for which one can prove that this process converges and produces the correct weights. Nonetheless, in most cases we have no guarantee that the process will converge, so deciding when to terminate the training phase and make do with a network that does not always produce the correct classification is of the utmost importance.

It is important to realize that in the learning stage the system can simply "memorize" the input rather than extracting features that characterize the input. There are existing networks, such as the Hopfield network discussed later, whose main task is to memorize. However, learning networks are judged by their ability to give correct answers to new inputs for which no solutions were given. For example, assume that the training set consists of two sets of binary strings of fixed length. One set contains strings all having an even number of 1's, and the other contains strings all having an odd number of 1's. Our goal is that, after the training phase on the training data including specific strings, the network will be capable of classifying correctly as many strings that were not part of the training set.

To determine how well a network has learned we define a **test set** containing input items not presented to the network during the training phase but for which the correct answer is known. The success (i.e., comparing the network's result with the correct answer) of the network in classifying these items is the metric for evaluating the success of the network.

4.4 THE PERCEPTRON

One of the simplest types of a neural network that has learning capabilities is called the **perceptron** and was among the first to be studied (Minsky and Papert, 1969). Perceptrons are used to solve classification problems—that is, given an input, the network has to determine the class to which the input belongs. We will present a **learning rule** used to update the weights as more classification samples are presented to the network, and we will prove that this rule allows the network to converge to correct weights, if a set of correct weights exists.

4.4.1 Definition of a Perceptron

In a **simple perceptron** the input cells are directly connected to the output cells as seen in Figure 4.4 (we will describe more complex architectures later). In Figure 4.4(a) a number of input cells (marked as X_1, $X_2,...,X_5$) are connected to a number of output cells (O_1,O_2,O_3). For simplicity we will usually consider perceptrons with a single output cell as seen in Figure 4.4(b). Such a network can obviously classify only into two sets, but it is not hard to extend the discussion to larger number of output neurons and a correspondingly larger number of classes since output neurons are independent. Note that in a network with a single output neuron, it suffices to index the weights using a single index, w_i.

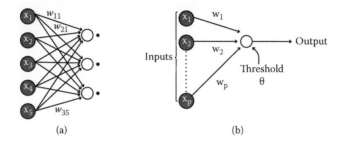

(a)   (b)

FIGURE 4.4   A simple perceptron. The general model (a) shows several inputs and outputs. However, since each output is independently calculated, we will analyze the behavior of a simple perceptron with a single output (b).

The Perceptron's Computation

The weighted input of the neuron in Figure 4.4(b) is the difference between the weighted sum of the incoming edges and the threshold value:

$$\sum_{i=1}^{p} w_i x_i - \theta$$

The value of the output neuron will be

$$O = \Theta\left(\sum_{i=1}^{p} w_i x_i - \theta\right)$$

where

$$\Theta(X) = \begin{cases} 1 & X \geq 0 \\ 0 & X < 0 \end{cases}$$

Note that a threshold level of θ can be achieved by a neuron with a threshold level of 0 by adding another input whose value is the constant −1 and the weight of the edge between it and the output neuron is θ. Adding the additional input without fixing the weight that connects it to the neuron allows the network to determine the threshold level according to the same learning procedure used for adjusting the network weights.

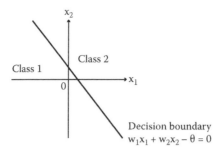

FIGURE 4.5   The weights of the perceptron define a line (in the two-dimensional case) that separates the data into two classes.

If we consider each input set as a point in an p-dimensional space, where each dimension represents one of the inputs $1,2,\ldots,p$, we see that the simple preceptron divides the space into two separate regions separated by a *hyperplane* of $p - 1$ dimensions defined by

$$\sum_{i=1}^{p} w_i x_i - \theta$$

In Figure 4.5 we can see how the network behaves with two inputs x_1 and x_2. If the point (x_1, x_2) is above the separating line then it belongs to Class 1; otherwise it belongs to Class 2. The boundary line is defined by $w_1 x_1 + w_2 x_2 - \theta = 0$. The slope of the boundary line is $-w_1/w_2$, and θ determines the distance from the origin. In particular, if $\theta = 0$, the boundary line passes through the origin.

In the two-dimensional (2-D) case depicted in Figure 4.5 the boundary is a one-dimensional (1-D) line. In the three-dimensional (3-D) case the boundary is a 2-D plane, and for more than three dimensions (i.e., more than three inputs), the boundary is a hyperplane. Note that the dimension of the hyperplane is always one less that the dimension of the input space. Separation by a hyperplane is called a **linear separation**.

Let us investigate using a simple perceptron to compute the values of Boolean functions, starting with the AND and OR functions. Their truth values appear in Table 4.1. If we consider X_1 and X_2 as coordinates on a two-dimensional plane and want to separate the points by their truth values we will get the situation depicted in Figure 4.6. This figure demonstrates the existence of a line separating the points with the

TABLE 4.1   Truth table of AND and OR

X_1	X_2	Result	X_1	X_2	Result
0	0	0	0	0	0
0	1	0	0	1	1
1	0	0	1	0	1
1	1	1	1	1	1
	AND			OR	

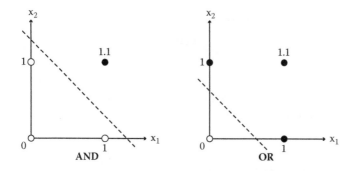

FIGURE 4.6   For the Boolean functions AND and OR, the coordinates (x_1, x_2) of each point represent the input values, and the color of the point represents the output (black circle for 1 and white circle for 0). It is easy to see that there are lines separating the 1's from the 0's.

value 0 from the points with the value 1. For instance, the line $x_1 + x_2 - 1.5$ (i.e., $w_1 = w_2 = 1$, $\theta = 1.5$) describes the function AND, and the line $x_1 + x_2 - 0.5$ describes the OR function. An infinite set of lines achieves each separation.

Consider now the XOR function, which outputs 1 if exactly one of its inputs is a 1. The truth table for XOR is given in Table 4.2, and Figure 4.7 shows it graphically. It is impossible to find a straight line with the black points on one side of it and the white points on the other side. In other words, the values of the XOR function cannot be linearly separated.

TABLE 4.2   Truth Table of XOR

X_1	X_2	Result
0	0	0
0	1	1
1	0	1
1	1	0
	XOR	

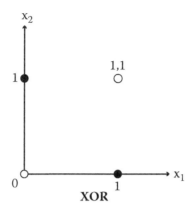

FIGURE 4.7   A simple perceptron cannot compute the XOR function. Note that it is impossible for a straight line to separate the 1 values (black circles) from the 0 values (the white circles).

4.4.2 Formal Description of the Behavior of a Perceptron

To make the description simpler we will use the *sign* function instead of the Θ function and represent the output by 1 and –1 instead of 1 and 0.

Let *p* be the number of input neurons in the network. We describe the weights of all the connections of the inputs to the output neuron as a *p*-dimensional vector **w**. Each input pattern **x**(i) will also be described as a *p*-dimensional vector (*i* is the ordinal number of the pattern). In vector notation, the value of the output neuron is

$$O = sign(\mathbf{w} \cdot \mathbf{x}(i))$$

where **w** · **x**(i) is the inner product of the two vectors,

$$\sum_{j=1}^{p} \mathbf{w}_j \mathbf{x}(i)_j$$

Recall that even though it would seem we have fixed the threshold value θ of the neuron to be 0, we can assume that one of the *p* inputs was included to take care of the threshold level.

Let O(i) be the correct output for **x**(i), that is, it is +1 if the input pattern belongs to the first class, and –1 if it belongs to the second class. Therefore, we are looking for weight vector **w** that will satisfy $O(i)=sign(\mathbf{w}\cdot\mathbf{x}(i))$.

The boundary between the positive and negative classes is the multi-dimensional plane described by the equation $\mathbf{w}\cdot\mathbf{x}(i) = 0$ which passes through the origin and is perpendicular to **w**. The product $\mathbf{w}\cdot\mathbf{x}(i)$ is equal to $\|\mathbf{w}\|\|\mathbf{x}(i)\|\cos\phi$ and has the value 0 when the two vectors are perpendicular, that is, when the angle ϕ between them is 90 degrees. As can be seen in Figure 4.8, the weight vector **w** has to be perpendicular to the separating line (or in the higher-dimensional case, the hyperplane) so that it can separate between the inputs whose output is +1 and those whose output is −1.

The geometrical meaning (Figure 4.8) is that one has to choose the weight vector **w** such that the projection of all input vectors **x**(i) on **w** has the same sign as $O(i)$.

Recall that the projection of the vector **x**(i) on **w** is:

$$\frac{\|\mathbf{w}\|\|\mathbf{x}(i)\|\cos\phi}{\|\mathbf{w}\|}$$

where ϕ is the angle between the vectors and $\|\mathbf{v}\|$ denotes the norm of the vector **v**. The *norm* or vector length is defined as

$$\|\mathbf{v}\|=\sqrt{\sum_{i=1}^{n}v_i^2}$$

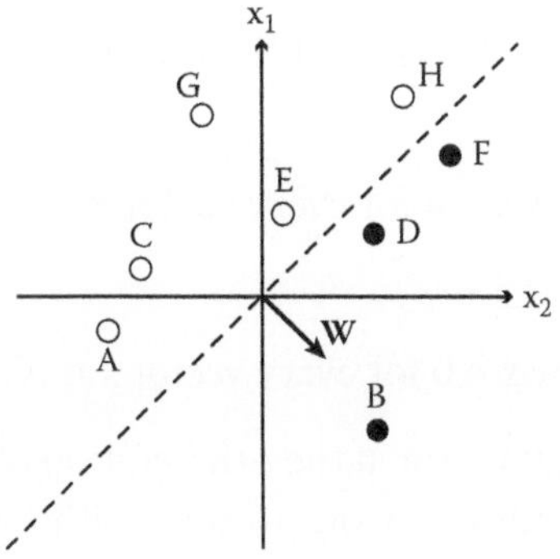

FIGURE 4.8 The weight vector **w** is perpendicular to the separation line between the two classes.

Note that even when the network has more than one output neuron, the computation for each output neuron is independent from the computations for the other output neurons (as each output neuron has a separate weight vector **w**), and therefore one can generalize the discussion of a perceptron with a single output neuron to perceptrons with a higher number of output neurons, such as the perceptron described in Figure 4.4(a).

4.4.3 The Perceptron Learning Rule

We have seen that if the input can be linearly separated a weights vector achieving the separation is present. In this section we limit our discussion to problems that are linearly separable and present a learning rule for a perceptron with a single output neuron, that is, a perceptron classifying the inputs into one of two possible classes. The rule will allow us to determine, iteratively, weights achieving the desired classification.

Since we are dealing with supervised learning, there is a training set containing samples for which the classification is known. Learning is achieved by feeding these examples to the network iteratively. If the network outputs the correct classification (where +1 represents the first class and –1 the second class), the weights are not changed. If the output is wrong, the weights leading from the inputs to the neuron are slightly modified to achieve the desired output.

As previously discussed we consider the threshold value θ as another input neuron with a value of –1, and for convenience this neuron will always appear first in the vectors, so all our input patterns will start with –1. The training set will be denoted by X. Let X' be the set of patterns belonging to the first class and X'' the set of patterns belonging to the second class ($X = X' \cup X''$). The goal of the training phase is to find a weight vector **w** such that

$$\mathbf{w} \cdot \mathbf{x} \geq 0 \text{ for every vector } \mathbf{x} \in X'$$

and

$$\mathbf{w} \cdot \mathbf{x} < 0 \text{ for every vector } \mathbf{x} \in X''$$

Let $\mathbf{w}(i)$ be the weight vector at the i-th iteration of training, and let $\mathbf{w}(0) = 0$. (In practice, it is advantageous to start with a weight vector consisting of small random weights rather than with an all 0 vector). We denote by $\mathbf{x}(i)$ the i-th training input pattern, and $t(i)$ is the target output (i.e., the required correct output) for this pattern.

The training is performed as follows for $i = 0,1,2,\ldots$:

1. Compute the output of the perceptron for pattern i:

$$\mathsf{o}(i) = sign\big(\mathbf{w}(i)\mathbf{x}(i)\big)$$

2. Determine the updated weight vector $\mathbf{w}(i+1)$, as follows:

$$\mathbf{w}(i+1) = \mathbf{w}(i) + \alpha\big[\mathsf{t}(i) - \mathsf{o}(i)\big]\mathbf{x}(i)$$

   where $0 \le \alpha \le 1$ is a constant that determines the learning rate.

3. Continue to pattern $i + 1$.

4. The training terminates if the weights are not updated during an entire epoch; otherwise, return to step 1 for another epoch.

When the pattern $\mathbf{x}(i)$ is classified correctly, the difference $\big[\mathsf{t}(i) - \mathsf{o}(i)\big]$ is 0, and therefore $\mathbf{w}(i + 1) = \mathbf{w}(i)$. If the pattern is misclassified, the weights are updated by adding or subtracting the quantity $\alpha\big[\mathsf{t}(i) - \mathsf{o}(i)\big]\mathbf{x}(i)$.

The learning rate constant α determines how big the weight changes are. If α is too large, the learning might be too "jagged": a low weight can increase to a value that will be too high and will misclassify the next pattern, and then the weight will decrease too sharply, causing more misclassifications, and so on. On the other hand, if α is too small, the training may be too slow and ineffective. There are no good rules for determining the value of α, and one has to resort to trial and error in each particular case.

To use a perceptron with more than one output neuron, the same training algorithm is applied to each output neuron separately. If the output is linearly separable for each output neuron, this will result in an appropriate set of weight vectors.

4.4.4 Proving the Convergence of the Perceptron Learning Algorithm

In the previous section we presented the algorithm for updating the weights of the perceptron. It is not clear that this algorithm always converges; in principle we can envision a situation in which the weights will be updated indefinitely, sometimes undershooting and sometimes overshooting the desired output. Fortunately, a mathematical proof is available to show that when the input is linearly separable the algorithm will converge. We will prove convergence for $\mathbf{w}(0) = 0$ and $\alpha = 1$.

The idea behind the proof is to follow the changes to the size of the weight vector $\mathbf{w}$ during the updates made by the perceptron learning rule. We will show that the size $\|\mathbf{w}\|$ of the weight vector grows faster than or equal to a term dependent on n^2, where n is the number of learning iterations (note that n may be larger than the number of patterns). On the other hand we will show that size of the weight vector grows slower than or equal to a term dependent on n. Since asymptotically a term dependent on n^2 will surpass a term dependent on n, there must exist an iteration where these two bounds meet, and the weight vector will not change further.

We note that the algorithm will behave identically for the problem of finding a vector $\mathbf{w}$ separating X' and X'' and the problem of finding a vector $\mathbf{w}$ that satisfies $\mathbf{w} \cdot \mathbf{x}(i) \geq 0$ for all the vectors $\mathbf{x}(i)$ belonging to the set consisting of X' and all the negatives (i.e., $-\mathbf{x}(i)$) of the vectors belonging to the set X''. Thus, we can assume for the proof that all the n inputs that were misclassified are misclassification of the form $\mathbf{w}(i) \cdot \mathbf{x}(i) < 0$ for $i = 1,2,\ldots,n$, where $\mathbf{x}(i) \in X'$.

Given that we have started with $\mathbf{w}(0) = 0$, according to the learning rule

$$\mathbf{w}(n + 1) = \mathbf{x}(1) + \mathbf{x}(2) + \ldots + \mathbf{x}(n) \tag{4.1}$$

Since we assume that the input is linearly separable, there exists a weights vector $\mathbf{w}^*$ such that $\mathbf{w}^* \cdot \mathbf{x} \geq 0$ for $\mathbf{x} \in X'$, and $\mathbf{w}^* \cdot \mathbf{x} < 0$ for $\mathbf{x} \in X''$.

Multiply equation (4.1) by $\mathbf{w}^*$ (the products are inner products):

$$\mathbf{w}^* \cdot \mathbf{w}(n + 1) = \mathbf{w}^* \cdot \mathbf{x}(1) + \mathbf{w}^* \cdot \mathbf{x}(2) + \ldots + \mathbf{w}^* \cdot \mathbf{x}(n) \tag{4.2}$$

Pick the minimal term on the right-hand side of (4.2) and denote it by p:

$$p = \min_{x(n) \in X'} \mathbf{w} \cdot \mathbf{x}(n)$$

So, from equation (4.2)

$$\mathbf{w}^* \cdot \mathbf{w}(n+1) \geq np$$

Squaring, we get

$$\left[\mathbf{w}^* \cdot \mathbf{w}(n+1)\right]^2 \geq n^2 p^2 \tag{4.3}$$

Recall that the product of the norms of two vectors is at least as large as their inner product (Cauchy–Schwartz inequality) and thus

$$||\mathbf{w}^*||^2 ||\mathbf{w}(n+1)||^2 \geq \left[\mathbf{w}^* \cdot \mathbf{w}(n+1)\right]^2 \tag{4.4}$$

Combining (4.3) and (4.4) we get

$$||\mathbf{w}^*||^2 ||\mathbf{w}(n+1)||^2 \geq n^2 p^2$$

or

$$||\mathbf{w}(n+1)||^2 \geq \frac{n^2 p^2}{||\mathbf{w}^*||^2} \tag{4.5}$$

Now we turn to show an upper bound on the growth of $\mathbf{w}$.
For every $k = 1, \ldots, n$

$$\mathbf{w}(k + 1) = \mathbf{w}(k) + \mathbf{x}(k)$$

Therefore, after taking the square of the Euclidian norm, we get

$$||\mathbf{w}(k+1)||^2 = ||\mathbf{w}(k)||^2 + ||\mathbf{x}(k)||^2 + 2\mathbf{w}(k) \cdot \mathbf{x}(k) \tag{4.6}$$

We assumed that the perceptron misclassified $\mathbf{x}(k)$, that is, $\mathbf{w}(k) \cdot \mathbf{x}(k) < 0$ and therefore (4.6) implies that:

$$||\mathbf{w}(k+1)||^2 \leq ||\mathbf{w}(k)||^2 + ||\mathbf{x}(k)||^2$$

or

$$||\mathbf{w}(k+1)||^2 - ||\mathbf{w}(k)||^2 \leq ||\mathbf{x}(k)||^2$$

If we sum these inequalities for $k = 1, \ldots, n$, assuming $\mathbf{w}(0) = 0$, we see that most terms of the left-hand side cancel each other, and we are left with

$$\|\mathbf{w}(n+1)\|^2 \leq \sum_{k=1}^{n} \|\mathbf{x}(k)\|^2 \tag{4.7}$$

Pick the maximal term of the summation on the right-hand side of (4.7) and denote it by q:

$$q = \max_{x(k) \in X'} \|x(k)\|^2 \tag{4.8}$$

From (4.7) and (4.8) we get

$$\|\mathbf{w}(n+1)\|^2 \leq nq \tag{4.9}$$

From (4.9) we see that the growth of the length of the vector $\mathbf{w}$ is bounded from above by a term linear in n. Recall from (4.5) that the length of the vector is bound from below by a term dependent on n^2. Since asymptotically a term dependent on n^2 will surpass a term dependent on n, it is clear that there exists an iteration n_{max} for which

$$\frac{n_{max}^2 p^2}{\|\mathbf{w}^*\|^2} = n_{max} q$$

or

$$n_{max} = \frac{q\|\mathbf{w}^*\|^2}{p^2}$$

Therefore, we have shown that for $\mathbf{w}(0) = 0$ and $\alpha = 1$, and under the assumption that there exists a solution vector $\mathbf{w}^*$ the learning algorithm has to terminate after at most n_{max} iterations.

4.5 LEARNING IN A MULTILAYERED NETWORK

4.5.1 The Backpropagation Algorithm

We turn now to study multilayer networks in which layers of internal neurons exist between the input and the output layers. Such layers are often called **hidden layers**, since they are not "visible" from outside the network.

We have seen the limited computational powers of the simple single-layered perceptron: it can classify only patterns that are linearly separable. In their book Minsky and Papert (1969) showed that multilayered neural networks can compute functions that cannot be computed by a single-layer preceptron. For instance, it is enough to add one hidden layer to compute the XOR function, which is not computable by a simple perceptron. Multilayered networks' greater computing power is an attractive property, but they are much harder to engineer. Designing a suitable layout for a multilayered network and finding the appropriate network weights can be difficult. Finding an efficient learning rule for updating the weights of a multilayered network automatically was thus an important goal, but finding such a learning rule proved to be a real challenge. The challenge was solved by **Rumelhart** et al. (1986), who proposed an algorithm called feed-forward–**backpropagation.** Feedforward is an obvious property of multi-layered networks where the computation results are fed forward layer after layer from the input layer to the hidden layers and from them to the output layer. The term backpropagation characterizes Rumelhart et al.'s learning procedure, which is based on percolating the weight updates from the output level back to the input layer. For short, this algorithm is often called backpropagation.

The algorithm is based on examining the error in the output neuron by comparing it with the target value. If the value is incorrect, and hence the network weights need to be adjusted, the adjustments are backpropagated from one intermediate level to the previous one. At every level, we can define an error function for each node and try to minimize it. This can be done by a **gradient descent algorithm.** This algorithm uses the derivatives of the error function to calculate the change in the direction and magnitude of the weight vector that will minimize the error.

Recall that the activation function Θ we used to define the behavior of the output neuron in the simple perceptron is a threshold function and as such is neither continuous nor differentiable. We would like to replace it with a differentiable function with similar characteristics, that is, a function that returns one value (e.g., 0) for inputs that are smaller than the threshold value and another value (e.g., 1) for inputs larger than the threshold value with the added stipulation that the transition between the two values is as abrupt as possible, yet continuous. A function with such properties is the sigmoid function:

$$y = f_{sigmoid}(x) = \frac{1}{1+e^{-x}}$$

Figure 4.9 plots the sigmoid in the range [−5,5], and it is easy to see that it has the desired properties

We could adjust its slope by looking at the more general sigmoid

$$y = \frac{1}{1+e^{-\beta x}}$$

where β is a parameter.

A useful property of the sigmoid function is that its derivative can easily be expressed using the value of the function itself. This property will be useful later on:

$$y' = \frac{1}{1+e^{-x}} \cdot \left(1 - \frac{1}{1+e^{-x}}\right) = y(1-y)$$

Other functions such as the hyperbolic tangent function

$$y = \frac{e^{2x}-1}{e^{2x}+1}$$

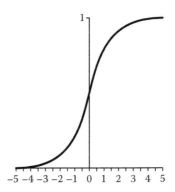

FIGURE 4.9   The sigmoid function has the property of a sharp transition between 0 and 1.

that returns values between –1 and +1 can be used, but the sigmoid function is most commonly used.

The output neurons will be denoted by the vector O (a single neuron will denoted by O_i). The output value of the hidden neurons in layer k will be denoted by V^k (and a single neuron by V_i^k), and the weight of the edge between neuron j in layer k–1 to neuron i in layer k will be denoted by $W_{i,j}^k$ (so the weights between the inputs and the first hidden layer are $W_{i,j}^1$). The vector X will denote the inputs. The values of the output neurons can be either binary or any real numbers. We assume N inputs, L patterns in the training set, and M layers (in addition to the input layer, designated as layer 0) in the network (Figure 4.10).

We set initial random values for all the weights and the network computes in a feedforward fashion similar to the operation of the single-layer perceptron, where each layer is evaluated after the evaluation of the previous layer except that the activation function is now the previously described sigmoid.

The input to a neuron V_i^k is

$$\sum_{\substack{\text{all edges } j \text{ that} \\ \text{enter the node } i}} W_{i,j}^k V_j^{k-1}$$

Therefore, the output value of that neuron will be

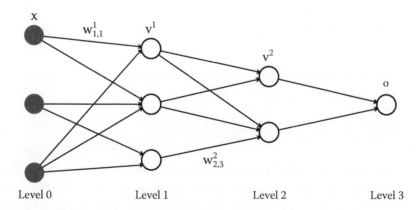

FIGURE 4.10 Artificial multilayer neural network.

$$V_i^k = f_{sigmoid}\left(\sum_{\substack{\text{all edges } j \text{ that} \\ \text{enter the node } i}} W_{i,j}^k V_j^{k-1}\right) = \frac{1}{1+\exp(-\sum_j w_{i,j}^k v_j^{k-1})}$$

While the computation of a multilayered network is quite similar to the computation of a simple perceptron (with the notable change of using the sigmoid function), the learning procedure (i.e., the way the weights are adjusted) is quite different.

The weights update stages are as follows:

1. Define the *error* function e for the output layer by

$$e_i^M = \frac{1}{2}(T_i - O_i)^2$$

where O_i is the actual output and T_i is the required output in the output i.

2. The derivative of the error function is used to compute the correction values that will be used to update the weights. Note that this involves taking partial derivatives, since the error function depends on the output of the nodes, which in turn depend on their input, which depends on the weights, and we are interested in the effect of each weight. As these derivations require application of the chain rule and are somewhat involved mathematically, we will not get into the details here and present only the final formulas, which are as follows:

    a. For the edges going into an output neuron compute the required corrections values as follows

$$\delta_i^M = (T_i - O_i)O_i(1 - O_i)$$

    b. To compute the correction values δ for previous layers, recall that the correction percolates down from the output layer; therefore, the correction for a neuron in a particular layer *m*–1 should take into account the corrections already calculated for neurons in layer *m*.

$$\delta_j^{m-1} = V_j^{m-1}\left(1-V_j^{m-1}\right) \sum_{\substack{\text{all edges } i \\ \text{that leave } j}} W_{i,j}^m \delta_i^m$$

for $m = M-1, M-2, \ldots, 1$.

An example of the calculation for a specific neuron is shown in Figure 4.11.

3. The needed changes to the weights are then

$$\Delta W_{i,j}^m = \alpha \delta_i^m V_j^{m-1}$$

Note that the correction for a weight is proportional to the activity of the previous node and the correction value reflecting the propagated error as computed by the formulas above. Like in the simple perceptron algorithm, α is a constant that determines the rate of learning.

4. The final update of all the weights in the network is simply

$$W_{i,j}^{new} = W_{i,j}^{old} + \Delta W_{i,j}$$

See an example in Figure 4.12.

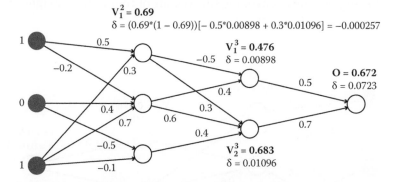

FIGURE 4.11   An example of the backpropagation algorithm. Assume that the output value should have been 1 but the network computed the value of 0.672. The algorithm calculates back (i.e., from the output layer at the right) the correction values δ for each node in the network according to the formulas given in Stage 3. In the example we detailed the computation of the correction needed for the top node in the first hidden layer, based on its output activity V and the δ values that propagate back from the two nodes in the second hidden layer.

5. Repeat all the previous steps for the next input pattern in the training set.

6. Repeat all the previous steps for the whole training set for the next epoch.

Unlike in the simple perceptron case, for the backpropagation algorithm we don't have guaranteed convergence. Thus, the obvious question is when we should stop training the network—or how many epochs to run. We can choose among several halting criteria: halt when the error on the training set is small enough; halt when the size of the weight updates for each epoch is small enough; halt when the network gives satisfactory results on the test set. The latter is usually the preferred criterion.

We have presented the simplest and most popular version of the backpropagation algorithm. This learning algorithm is probably the most common and most studied algorithm in the field of artificial neural networks. Thus, many variations of this basic algorithm have been suggested over the years such as using a method different from gradient descent for determining the weights or changing the parameters of the algorithm during the run—for example, accelerating the rate of weight change for edges whose weights has grown consistently over the previous iterations and slowing the rate of changing the weights of edges whose weights have oscillated during the last iterations.

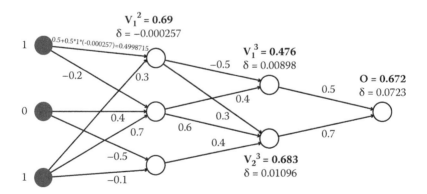

FIGURE 4.12 Updating the value of the weight between the top input node to the top node of the hidden layer according to the formula given in Stage 5 when $\alpha = 0.5$.

As with other optimization algorithms (e.g., genetic algorithms), we have to consider the possibility of hitting a local minimum (or maximum) when using backpropagation, a situation where the weights cannot be changed further by applying the learning algorithm even though a better weight vector exists. If we focus on the error function of the output neuron, we can think of these situations as being trapped in a local minimum of the output neuron error function instead of finding the global minimum for which the learning algorithm searches.

Several techniques can be used to address this problem. One reason for local minima might be choosing bad initial weights. If they are too large, the function will tend to consistently overshoot and produce a value of 1, which may make training difficult. Therefore, it is advantageous to choose initial weights such that a typical neural input will have a value smaller than 1.

An important approach for avoiding local minima is to increase randomness. This increases the region of all possible weights sampled during training and can be achieved in different ways:

- Train the network by selecting randomly the order in which the input patterns from the training set are presented to the network. This decreases the probability that patterns close to each other in the training set will cancel each other's influence.

- Modify the algorithm so that during training it will occasionally (but rarely) change the weights to increase the error function. While it seems that this would decrease the probability that the network will converge to the correct values, it turns out that this technique can prevent the algorithm from being trapped in a local minimum. Clearly, such uphill moves should be performed infrequently. It is reasonable to reduce the probability of performing this operation as the learning progresses to minimize the damage such moves make to the learning that has been already achieved.

- Add **noise** to the system. In this method, the weights are occasionally changed randomly. The random change is small to avoid disturbing the learning that has already been achieved. On the other hand, a small change in weights may be enough to move away from a local minimum. Another way of adding noise is randomly changing the value of the input patterns presented to the network. Experience shows that adding noise in either way can improve the ability of the

network to **generalize** (i.e., produce good results on the patterns in
the test set).

4.5.2 Analysis of Learning Algorithms

It is important to note that learning is an unusual algorithmic problem. It
may even be viewed as an ill-defined problem. The reason is that the input
(i.e., the training set) does not necessarily define a single solution. Often
many solutions perform well on the training set, but it is impossible to
forecast how well they will do on the test set and in the real world. The
goal of learning is to find a model that will represent the input data well,
but many such models may perform well on the training set. Different
learning algorithms will give rise to different results for the same training
set in accordance with the properties of the algorithms.

It is useful to consider two types of learning errors: (1) **training
errors**, which are errors in dealing with patterns in the training set;
and (2) **generalization errors**, which are errors in dealing with new
patterns from the test set. Despite the fact that the goal is to minimize
both kinds of errors, it turns out that minimizing one type can increase
the prevalence of the other type, so in fact one has to find a way to balance these goals. The reason for this trade-off is that one can build a
model that fits the training set members well, at the cost of creating
a model that is very specific to this set and does not perform well on
new samples. In general, at the beginning of the learning process both
the training error and the generalization error will decrease, but after
a certain point, as members from the learning set are learned again
and again the network performs better on them, thereby decreasing the
training error but at the price of increasing the generalization error as
can be seen in Figure 4.13.

A simple example of this situation is finding a curve to fit a dataset.
Look at the two situations depicted in Figure 4.14. Both graphs attempt to
describe the four points, but while the left curve passes exactly through
the points, the right one does not pass through any of them. Nonetheless,
it is reasonable to believe that it will model better new data points. On the
other hand, while a straight line seems to be a good approximation for the
current points, we do not know if the sample set is a good representative
of the data, and introducing more points might lead us to prefer a more
complex model that is not a straight line.

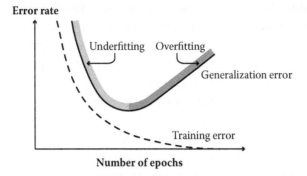

FIGURE 4.13 Increasing the length of training improves the performance on the training set but may adversely affect the performance on the test set as the network starts to overfit the data.

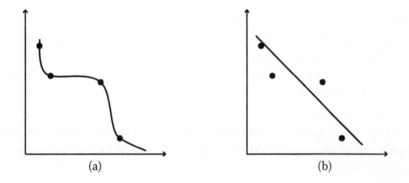

FIGURE 4.14 Two types of lines that describe the given data points. The line on the left (a) goes through all the points, and the line on the right (b) does not go through any of them. Nevertheless, in many situations the description provided by the straight line on the right is more informative as it captures the trend in the data and allows a better prediction of the value of additional points.

In general, the closer we get to the desired answers for the training set, the larger the danger of **overfitting**, that is, a situation where the model describes the training set exactly, yet lacks generality (as is the case in the left panel of Figure 4.14). The opposite case where the model contains too little data to describe the training set is called **underfitting**. Due to the danger of overfitting it is typical not to continue to train the network until it is very successful on the training set but to stop training before that to increase the chances of obtaining good results on the test set.

Increasing the number of neurons in the network (and in particular the number of neurons in the hidden layers) can also cause overfitting. This puts the designer of the network in a delicate position: the network cannot be too small, since this may cause underfitting, or too large, in which case it can learn the training set too well, leading to overfitting. Adding noise, which we previously described as a way to avoid the convergence of the weights to local minima, can also be useful tool to reduce the risk of overfitting. For example, one can, on occasion, supply the learning process with a wrong classification for a member of the training set. By reducing the probability of converging to an exact solution for the training set, this seemingly strange tactic can actually improve generalization.

4.5.3 Network Design

So far we have assumed that the architecture of a network is given and the goal of the learning process was to determine the weights of the edges. An obvious question is how to design the network—that is, how to decide how many hidden layers it will have and which pairs of neurons to connect with one another. Experience shows that usually it is not beneficial to deal with networks with too many layers of hidden neurons, and in fact that might harm performance. In practice it seems that two or three layers suffice for most problems that can be solved by the backpropagation method.

As for connectivity, recall that in these networks neurons can communicate only if they are in consecutive layers, but it is not always necessary to connect every neuron in one layer with every neuron in the consecutive layer. Obviously, the more connections, the more weights have to be determined, a situation which may hinder the network's ability to learn in a reasonable time and may also cause overfitting. One way of dealing with this situation is to train the network for a relatively short period of time and then to sever the connections that have weights close to 0. Next, the network is retrained, and the other weights are updated. This process can be repeated until the network has a "reasonable" number of edges.

Another way to design a network is by using an external algorithm that will design the network optimally. For instance, one could use a genetic algorithm that deals with a population of neural network of different architectures. At every generation of the genetic algorithm, the performance of the neural nets will be assessed (e.g., by testing how well they do on the test set), so that selection will prefer the more successful networks.

Another important question is that of **data representation**. The network designers have to decide how to represent the data as an input to the network, and even more importantly they have to decide which data to present the network so that it succeeds in learning. This is true for both the input and the output. This is similar to the situation we encountered in the discussion of genetic algorithms where it is important to consider how to represent each solution as a chromosome. In both cases there are no recipes for solving these issues, and the designer has to rely on experience and understanding of the problem domain.

We will start with a simple example. Assume that we want to classify a set of data points into n subclasses. In principle we can use $\log_2 n$ output bits to represent the output. Thus, if we want to classify data into eight subclasses, we can use binary representation of three bits. However, this requires the network to "learn" the binary code in addition to solving the particular classification problem it encounters. Often, especially when the number of input examples in not very large, this is beyond the capabilities of the network. Thus, when the goal is to classify a dataset into n subsets, it is better to work with n output neurons and a unary representation, so that if neuron k has the value 1 and all other $n-1$ neurons have the value 0, the result of the classification is k.

Another more fundamental example for the importance of representation comes from image processing. If the image consists of 1000 × 1000 24-bit values pixels, it would seem one should use a network with 24,000,000 inputs. This is often not practical because of the size of the network and the time required to learn the weights. Therefore, it is reasonable to preprocess the data and try to extract features required for the computation. For example, one can locate lines, corners, and changes in the density of color. Actually, this is similar to the way the human brain which contains a large set of sensors that extract such features operates (Marr, 1982). It turns out that the retina contains cells whose electrical activity depends on identifying a particular pattern of boundaries between light and darkness; that is, some cells are sensitive to lines and react only when the field of vision contains lines but not surfaces of uniform color. Similarly, some neurons are sensitive to the movement of objects in the field of vision and can even distinguish the direction of motion. Such biological mechanisms suggest that similar ideas should be used by artificial neural networks as well.

4.5.4 Examples of Applications

We now present a few classic examples of using neural networks. Most of these are of multilayered networks trained using backpropagation. Nonetheless, in many of the examples we discuss, some deviations from the standard algorithm were used to adapt the network to the requirements of the problem.

NetTalk

The goal of this project (Sejnowski and Rosenberg, 1987) was to build a network that translates written text into speech. The input was seven characters from the text, and the network had to determine the pronunciation of the middle character. The seven-character window through the text is used to allow the network to determine the pronunciation based on the context of each character, as seen in Figure 4.15. Two forms of data were considered. In the first, the text in fact came from a transcript of recorded continuous speech that was moved along the window such that more than one word could be included at the same time in the window (as can be seen in Figure 4.15). In the second dataset, dictionary words were fed into the system one at a time.

For every seven characters, the network determined what sound (**phoneme**) to make out of a set of 26 possible phonemes. The network contained $29 \times 7 = 203$ inputs: seven characters in the sliding window, where each character may be one of the 26 letters of the language and three punctuation marks; each character is represented by 29 neurons, one with the value 1 and all other neurons set to 0. The network contained 80 hidden neurons in one layer and 26 output neurons that specify the output sound. After 50 passes over the training set the network was 95% and 98% precise in its output for the first and second datasets, respectively.

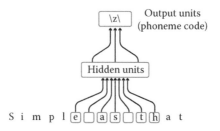

FIGURE 4.15 A schematic representation of the network used by NetTalk to identify phonemes in a "text to speech" translation system.

The self-organization of the network was quite interesting. It turns out that the network learned how to identify certain components of the input. It began by learning very general properties (e.g., the transition from word to word), and progressed to more subtle properties. After training, some of the hidden neurons were capable of identifying well-known properties, such as the distinction between vowels and consonants. The network was 78% accurate on new text (generalization)—a result that, while not very high, was still high enough to make the speech understandable.

Another interesting test of the network was determining its fault tolerance. We discussed already that fault tolerance is one of the main differences between the brain and digital computers. And indeed, in this case a local fault in the network (the removal of some neurons) or a random change in weights caused a gradual degradation in the quality of the output rather than a catastrophic complete failure. Moreover, after such failures, retraining quickly enabled the network to recover.

Handwriting Recognition

In this example, a neural network was used for automatically reading zip codes off envelopes, as part of a system to automatically sort mail for the U.S. Postal Service (LeCun et al., 1989). The system made use of many technologies, but we will discuss only the aspects related to the topics discussed in this chapter.

The training set consisted of about 10,000 handwritten digits. Finding the zip code on the envelope and separating it into distinct digits are complex tasks that will not be discussed here. The task we will discuss is the final stage of the process: identifying each separate digit.

The input to the neural network was a 16×16 matrix of pixels representing the image of a digit. The characters were scaled as needed so that they were all the same size, regardless of the particular handwriting. Every pixel in the matrix was represented by an input neuron. The values of the inputs were continuous rather than binary (black or white) because the scaling may cause every pixel in the matrix to represent more than one pixel in the original image. Figure 4.16 shows examples of the kind of handwritten digits presented to the neural network.

The network contained three hidden layers and an output layer containing 10 neurons representing the digits 0–9. The output of the network was determined by the output neuron with the highest value. The organization of the network can be seen in Figure 4.17.

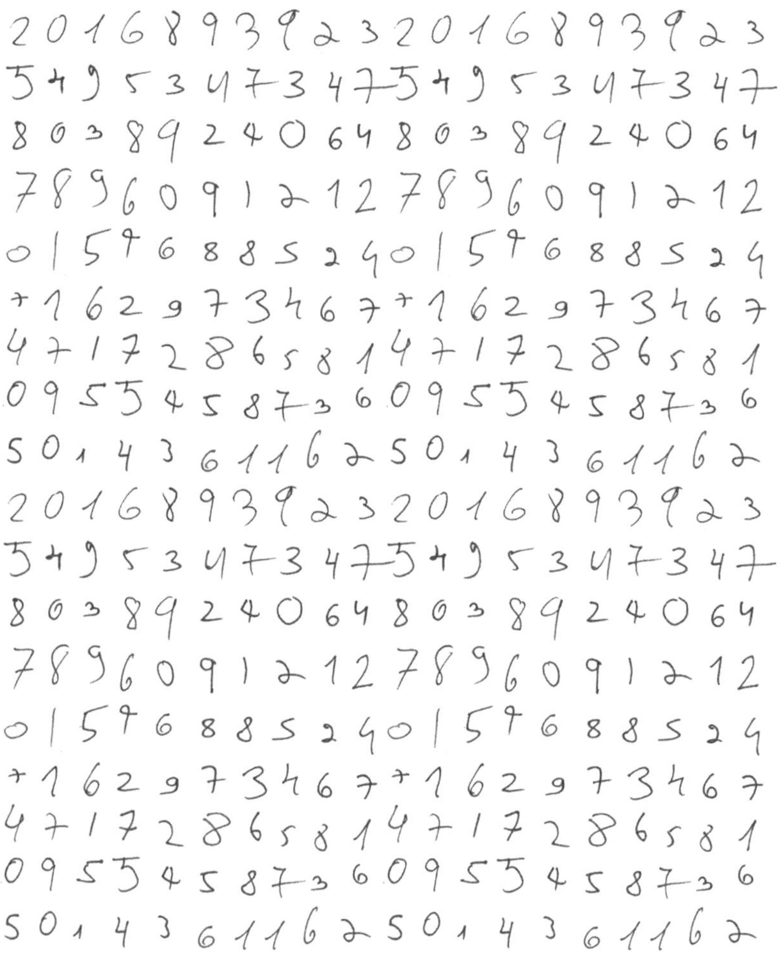

FIGURE 4.16 Example of handwritten digits. The task of the neural network is to identify such digits in spite of the large variability in how people are used to writing the different digits.

The first two hidden layers are used to identify recurring features in the input images:

- The first hidden level contains 12 groups of 8 × 8 = 64 neurons each, which represent an 8 × 8 matrix tasked with identifying a specific feature. The input for each neuron in each group is a 5 × 5 neighborhood on the input matrix; the corresponding neurons in each of the feature detectors "observe" the same neighborhood on the input matrix. Adjacent neurons deal with neighborhoods that are two pixels

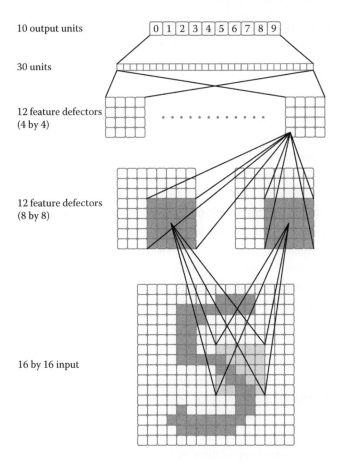

10 output units

30 units

12 feature defectors
(4 by 4)

12 feature defectors
(8 by 8)

16 by 16 input

FIGURE 4.17 A schematic representation of the neural network's architecture used to identify handwritten digits. (Adapted from LeCun, Yann, et al., *IEEE Communications Magazine* 27, no. 11, 41–46, 1989. With permission.)

apart in the input matrix. All the neurons in each group have the same weights. (Interestingly, even though the weights are the same, the threshold could differ from neuron to neuron.) Functionally, this means that all neurons in the group identify the same image feature, but in different regions of the input. This **weight-sharing** property decreases the number of weights the network has to learn, allowing the training set to be smaller.

- The second hidden layer is quite similar to the first layer. It contains 12 groups of $4 \times 4 = 16$ neurons that observe a region of 5×5 cells in the previous layer. The neurons receive their inputs from corresponding regions from 8 of the 12 groups in the first layer. (Every group of 4×4 neurons observes a different combination of eight groups.)

- The third hidden layer contains 30 neurons, all of which are connected to all the neurons in the previous layer.

- The output layer contains 10 neurons representing each of the digits, all of which are connected to all the neurons in the previous layer. The identity of the input digit is determined by the identity of the most active output neuron.

The network contained 1256 neurons, 64,600 edges and 9760 weights that had to be learned. The network was trained on 7300 digits and tested on 2000 images. The error rate was 1% on the training set and 5% on the test set. The latter could be reduced to 1% by rejecting cases where the value of the most active neuron is very close to that of the runner-up. These cases would then have to be read by a human operator. This solution gave rise to 12% of the test inputs being rejected.

After training, the system was implemented on standard signal processing hardware. The resulting system was capable of processing more than 10 digits per second (from photographing the image to identifying the digit by the neural network).

Backgammon

To play well, a computer system must be able to determine which of the next possible moves in a game is the most advantageous. The ranking of moves is usually based on two elements: (1) a good function for static evaluation of a given configuration of the board; and (2) an algorithm that can simulate the progression of the game a few moves ahead and can decide which move to select such that the position of the board will be, after the next few moves, the best possible for the player the computer represents (the minimax algorithm is an example of how this idea is implemented). Writing good backgammon-playing software poses a computational challenge since the number of possible moves after the die are tossed is usually rather large (about 20 moves on average), and there are 21 possible outcomes of the die toss. Therefore, the space of possible moves grows exponentially fast—faster than chess.

The approach we mention here (Tesauro and Sejnowski, 1989) does not explicitly look ahead: Given a position on the board and the dice value, the task of the network is to judge the quality of the possible moves.

A neural network was used to grade each possible move. The network was trained to grade each triplet of the from {current board configuration, dice values, possible move} with values between –100 (bad) and 100 (excellent). The network was trained using about 3000 such triplets, where the grading provided by the training set was determined by a skilled backgammon player. The inputs were the triplets and a few specific properties calculated from the state of the board (e.g., the number of blots, which are single checkers that can be hit). Altogether, 459 neurons were needed to represent the input. The network contained two hidden layers of 24 neurons each and one output neuron whose value was a real number between 0 and 1 that could be transformed to values in the range of –100 to 100 used to grade moves.

An important element in training the network was the use of noisy input. The input contained a few moves whose grades were determined randomly to reduce the danger of overfitting. This is critical for backgammon as the number of possible backgammon moves is much larger than the size of any reasonable training set. Cases where the network graded an input particularly badly were identified and rectified by adding correcting samples to the training set with the aim of reducing the occurrence of similar problems.

When the trained network played against another computer program (not implemented using a neural network), it won in 59% of the games. Interestingly, when only the triplets of {current board, dice, move} were used as input to the network, without the additional information calculated from the board, the machine won only 41% of the games. This significant drop in performance highlights the importance of preprocessing of the data presented to neural networks, since all of the additional features that were used could in principle be extracted automatically from the board, but apparently the system was not powerful enough to find them. In addition, it was noticed that, if no noisy data were included in the training set, the success rate dropped from 59% to 45%, indicating that noisy data are indeed useful in preventing overfitting.

The Bottleneck Technique

Neural networks and backpropagation can be used for finding efficient representations for large sets of patterns and for identifying common properties. This can be achieved by training a network with N input and N output neurons and only one hidden layer with M neurons where M is significantly smaller than N. The goal of training is to find weights for the hidden neurons so that each input pattern presented to the network will be reproduced as output. At first glance this may seem a strange task. Why should we want to reproduce our input? The answer lies in the observation that the hidden layer has to represent the input pattern so the output layer can reproduce it. Thus, if M is smaller than N and the training is successful, then we have actually found a way of representing the N input bits of information by a smaller set of M bits. In other words, if we have "squeezed" the data through a narrow bottleneck and were able to reproduce it back, then the hidden layer must have found properties and interrelationships in the dataset. This technique can be used for text or image compression or more generally for identifying features of a dataset.

4.6 ASSOCIATIVE MEMORY

4.6.1 Biological Memory

Our discussion so far has centered on using the computational capabilities of neural networks for solving problems such as classification. The human brain (as well as the brains of other species from mollusks to elephants) has another important role, namely, memory. Brains can remember and retrieve a huge number of data items. Despite the fact that conventional computers have very large memories, their memory capabilities are very different from those of animals. After many years of research and many insights that have been gained, many aspects of the memory capabilities of the brain are still not well understood. We will see how to implement some of the biological principles in a simple neural network architecture.

As far as we know, in the brain there is no explicit distinction between the neurons used for computations and those used for memory. We know of no type of neurons whose specific function is memory. Nonetheless, certain brain regions are dedicated to memorizing, and patients who suffer from damage (due to injury or illness) to these regions experience various types of memory loss while other cognitive functions are not impaired. It is common to classify memory to short-term memory (a few seconds long), which can store a very limited amount of data (the common claim is that it can store up to about seven data items), and long-term memory, which can potentially be retained for a lifetime (though obviously not all long-term memories are retained).

It is common to discuss three facets of the memory system: (1) **encoding**; (2) **storage**; and (3) **retrieval**. Here, we will not discuss the encoding (i.e., the way the information from the senses is converted into neural information that can be further analyzed by the brain), since it is specific to the biology of the brain and is less relevant to our discussion of artificial neural networks.

The brain's storage system is distributed; that is, no single cell stores a particular data item. A centralized arrangement would have caused memories to disappear with the natural death of the individual cell storing a particular datum of information. Thus, brains must have mechanisms to ensure robustness. One way of achieving robustness is to keep multiple complete copies of the data in different locations (similar to computer backups). However, evidence suggests that memories are stored in the brain in a distributed fashion. For example, data items seem to be broken up into smaller units that can be reassembled, even if a few of the pieces are missing.

From daily experience we know that memories are often retrieved by association. For instance, we recall a person's name when seeing his or her face, or a tune gives rise to memories of an event during which the tune was played. The retrieval can be instantaneous and not conscious, but occasionally we have to explicitly search our databanks to recall a data item. Associative memory allows us to recall data not by accessing a particular location in memory (the way digital computers deal with memory) but rather by using partial content to access the rest of the data. This property is called **content addressability**.

4.6.2 Hopfield Networks

In 1982, **John Hopfield** described an artificial neural network that provides associative memory. Rather than being organized in layers like the networks we discussed in previous sections, each neuron in a Hopfield network is connected to all other neurons. Moreover, there is no explicit distinction between input and output neurons. We will see how the network remembers by updating the weights and how data retrieval is executed by an iterative process of updating the values of neurons connected by these weights. As no neurons represent the output, retrieval ends when all the neurons reach a steady state where their values no longer change.

4.6.3 Memorization in a Hopfield Network

Let us start by defining the storage process and understanding the retrieval process. Assume we have to remember p strings U of N bits each. Let U_i^k

be the i-th bit of the k-th string. The values of the memory table will be computed as follows:

$$T_{i,j} = \sum_{k=1}^{p} \delta(U_i^k, U_j^k) \tag{4.10}$$

where

$$\delta(U_i^k, U_j^k) = \begin{cases} 1 & U_i^k = U_j^k \\ -1 & U_i^k \neq U_j^k \end{cases}$$

Formula (4.10) allows us to compute the weights for a network used to store a set of samples. Each neuron in the network corresponds to one of the p bits. The value $T_{i,j}$ represents the strength of the connection between the two neurons. If we are supplied with the set of samples ahead of time, we can use formula (4.10) to compute the weights. Alternatively, the same value can be computed iteratively by considering sequentially each sample and updating the strength of the connections as required.

Let us look at the following example. Suppose the goal is to memorize five binary strings of length six: 001010, 111100, 101110, 010001, 011000 (Figure 4.18).

The strings will be stored in "associative memory," which is a 6 × 6 matrix (6 is the length of the strings), where the value in cell (i,j) reflects in how many of the strings bit i and bit j are identical (Figure 4.19). For instance, the value of cell (1,2) reflects the fact that the first and second bits are equal in two of the five samples and not equal in the other three, and therefore its value is 2 − 3 = −1. The rest of the table is filled up in a similar way. By definition the diagonal cells (i,i) are set to zero. The matrix

0	0	1	0	1	0
1	1	1	1	0	0
1	0	1	1	1	0
0	1	0	0	0	1
0	1	1	0	0	0

FIGURE 4.18 An example of input strings presented in the Hopfield network.

	1	2	3	4	5	6
1	0	−1	1	5	1	−1
2	−1	0	−1	−1	−5	−1
3	1	−1	0	1	1	−5
4	5	−1	1	0	1	−1
5	1	−5	1	1	0	−1
6	−1	−1	−5	−1	−1	0

FIGURE 4.19   The weight matrix that is calculated for the input strings shown in Figure 4.18.

represents the data in a distributed fashion since each cell contains information which is affected by all the data samples. Note that the learning rule (4.10) ensures that the weights between neurons that typically have the same value will be high.

This learning procedure is called **Hebb's rule,** named after the psychologist Donald Hebb who postulated that in the brain the connection strength between cells is correlated with the frequency of them being active together. This idea is often summarized by the slogan "Neurons that fire together wire together." Obviously, the simple learning rule in (4.10) cannot adequately model the complicated process happening in the brain (which, moreover, is not well understood). For instance, in our model two inactive cells (bits with value 0) will have a strong connection, which is probably not the case in a biological system.

4.6.4 Data Retrieval in a Hopfield Network

Retrieval is initiated by setting the value of the neurons in the network to the values of the corresponding bits in the string. The retrieval process should be such that, if the system is presented with one of the samples in memory the values of the neurons in the network will not be altered. If a slight variation is presented (i.e., a string where only one or two bits have been altered relative to the original), we want the network to come up with the appropriate (i.e., most similar) sample string. If an entirely new sample is presented, we usually have no expectations from the system.

The retrieval of the memorized patterns is based on the fact that the weight matrix represents the strength of the connection between all pairs of bits. Thus, when we want to retrieve the value of a specific bit, we can look at the values of the other bits and see what their "recommendation" is. For example, if the values of other bits that are strongly coupled with the given bit is 1, then we should set its value to 1. The actual calculation is achieved by the formula given in (4.11).

$$U_i^{NEW} = \begin{cases} 1 & \sum_j U_j T_{i,j} \geq 0, \\ 0 & \sum_j U_j T_{i,j} < 0, \end{cases} \qquad (4.11)$$

We will now repeat the process for other randomly chosen neurons until the process stabilizes, that is, until no neuron is updated. Intuitively, at every step the chosen neuron "adopts" the value "recommended" by the other bits. For instance, consider the string 1 1 1 1 1 0 (which is similar to the string 1 1 1 1 0 0 in Figure 4.18). Assume that we want to compute the value of the fifth bit:

$$SUM_5 = U_1 \times T_{5,1} + U_2 \times T_{5,2} + U_3 \times T_{5,3} + U_4 \times T_{5,4} + U_5 \times T_{5,5} + U_6 \times T_{5,6} =$$

$$1 \times 1 + 1 \times (-5) + 1 \times 1 + 1 \times 1 + 1 \times 0 + 0 \times (-1) = -2$$

Therefore

$$U_5^{NEW} = 0$$

This correction is indeed what we are hoping for, but it is only a step in the process. In general, we have to show that the process does indeed converge; that is, no infinite loop can arise where the values of a cell will oscillate between 0 and 1, and the process will eventually halt and retrieve the correct value.

Note that the process we have described is asynchronous, where at each time unit one cell is updated and its new value will be used to update the values of cells updated in subsequent time steps. This property is believed to be relevant to the situation in the brain where there is no master clock; therefore, the activity is not tightly synchronized. However, in principle one can also explore synchronous systems where all the values are changed simultaneously (similar to the synchronous updates in cellular automata discussed in Chapter 2). Clearly, a synchronous system would require auxiliary memory to prevent updating bits before they affected the other bits.

4.6.5 The Convergence of the Process of Updating the Neurons

We will prove the convergence of the process of updating the values of the neurons by defining a quantity called the total energy of the network:

$$E = -\frac{\sum_j \sum_{i(i \neq j)} T_{i,j} U_j U_i}{2}$$

We prove that the total energy is monotonic decreasing as the weights are updated. As the energy is additive, we can consider its component derived from bit j:

$$E_j = -\frac{1}{2}\sum_{i \neq j} T_{i,j} U_j U_i = -\frac{1}{2} U_j \sum_{i \neq j} T_{i,j} U_i$$

When the value of U_j changes, all the other bits remain as they were, so the difference in energy can come only from change in U_j. Therefore,

$$\Delta E_j = E_j^{new} - E_j^{old} = -\frac{1}{2}\Delta U_j \sum_{i \neq j} T_{i,j} U_i$$

Note that if there was no correction to the value of U_j then $\Delta U_j = 0$.

Otherwise, the value of U_j has been changed by the update rule in one of the following two cases:

$$\text{if } \sum_{i \neq j} T_{ij} U_i \geq 0 \text{ then } \Delta U_j \geq 0$$

$$\text{and if } \sum_{i \neq j} T_{i,j} U_i < 0 \text{ then } \Delta U_j < 0$$

In either case the product

$$\Delta U_j \sum_{i \neq j} T_{ij} U_i$$

will be positive; therefore, the total energy will decrease. As the system is finite, the energy cannot decrease ad infinitum, so the network has to converge. Next, we need to show that the values the network converges on will be those strings the network is tasked with storing.

4.6.6 Analyzing the Capacity of a Hopfield Network

Why and under what conditions will the system give the expected results and be able to retrieve the input? We begin by examining a network designed to store a single input sample. In this case the weights of the network will be

$$T_{i,j} = \delta(U_i, U_j) = \begin{cases} 1 & U_i = U_j \\ -1 & U_i \neq U_j \end{cases}$$

Applying the neuron update rule and computing

$$\sum_{j \neq i} U_j T_{i,j}$$

for every bit in the sample does not cause any updates.

In fact, the weights define an energy surface as depicted in Figure 4.20 where the saved string is an attractor for the network such that points close to the attractor will converge to the attractor. Obviously, if we start at the attractor (i.e., present a memorized sample to the network), the system will remain at the same point. Notice that, because of the symmetry between the 0 and 1 bits, the strength of all the connections will not change if we replace every 0 with 1 and every 1 with 0. Therefore, when one pattern is memorized its complement is memorized too, as can be seen in Figure 4.20.

What happens when more than one string is to be memorized? The energy surface will be much more complex and may contain multiple minima. Intuitively, there must be a limit to the capacity of the network, so that if we attempt to memorize too many samples the attractors will overlap and patterns will be attracted to the wrong attractors.

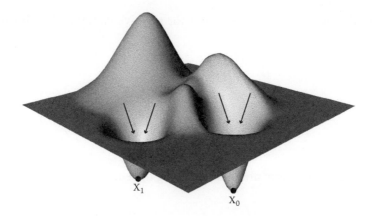

FIGURE 4.20   The energy surface of a Hopfield network that stores a single input. The energy surface (reflecting the energy function) is shaped such that the input data point will be at the minimum of the surface (X_0) and nearby points will be attracted to the minimum. Note that for binary strings the symmetrical treatment of 1 and 0 bits results in a "shadow" minimum (X_1) corresponding to a string where the bits are flipped.

Let us determine how many patterns can be memorized such that small perturbations will be corrected by the dynamics of the network. To make the analysis more convenient we will discuss neurons with the values +1 and –1 rather than 0 and 1, so the sign of the product of two bits indicates whether the bits are equal (when their product is 1) or different (when their product is –1). Moreover, we will normalize the weights in (4.10) by multiplying by $1/N$, where N is the number of bits in a string:

$$T_{i,j} = \frac{1}{N} \sum_{k=1}^{p} U_i^k U_j^k \qquad (4.12)$$

Note that by this definition the main diagonal of the weight matrix is 1 and not 0 and that the normalization achieved by multiplying by the constant $1/N$ may affect the retrieval rate but not its result.

We now consider one pattern U^v and check if it is stable. It will be stable if no neuron changes due to the update rule; that is, for every i it holds that

$$U_i^v = \text{sign}\left(\sum_j T_{i,j} U_j^v \right)$$

where the *sign()* function returns 1 for positive values and −1 for negative values.

By inserting the expression for $T_{i,j}$:

$$U_i^v = \text{sign}\left(\sum_j T_{i,j} U_j^v\right) = \text{sign}\left(\frac{1}{N}\sum_j\left(\sum_k U_i^k U_j^k\right)U_j^v\right)$$

Isolating the $k = v$ term and manipulating the equation gives

$$U_i^v = \text{sign}\left(\frac{1}{N}\sum_j\left(\sum_{k\neq v} U_i^k U_j^k + U_i^v U_j^v\right)U_j^v\right)$$

$$U_i^v = \text{sign}\left(\frac{1}{N}\sum_j\left(\sum_{k\neq v} U_i^k U_j^k\right)U_j^v + \frac{1}{N}\sum_j\left(U_i^v U_j^v\right)U_j^v\right)$$

$$U_i^v = \text{sign}\left(U_i^v + \frac{1}{N}\sum_j\left(\sum_{k\neq v} U_i^k U_j^k\right)U_j^v\right) \tag{4.13}$$

If the second term is 0, U^v is clearly stable. It will also be stable if the second term is small enough: if U_i^v is 1, then if the second term is greater than or equal to −1 it cannot flip the sign of U_i^v. Similarly, if U_i^v is −1 then the second term will have to be greater than 1 to change the sign. Since the sum of random (+1/−1) bits will tend to be around 0, in most cases the sum will not be larger than N and the pattern will be stable, but the probability of this depends on the number of strings p and their length N.

Let us determine the probability that

$$\frac{1}{N}\sum_j\left(\sum_{k\neq v} U_i^k U_j^k\right)U_j^v < -1$$

since in this case the sign of the expression in (4.13) will change. As a first approximation, let us assume that the patterns and the weights

TABLE 4.3   Error Probabilities and Network Capacity

$\dfrac{P_{max}}{N}$	P_{error}
0.105	0.001
0.138	0.0036
0.185	0.01 (= 1%)
0.37	0.05 (= 5%)
0.61	0.1 (= 10%)

are random, so the question boils down to the question of what is the probability that the product of $1/N$ and the sum of Np random numbers whose values are –1 or +1 is less than –1. In other words, what is the probability that sum of Np random numbers whose values are $-1/N$ or $+1/N$ is less than –1?

Assuming that p and N are large, by the central limit theorem this sum is distributed normally with a mean of 0 and variance of p/N, and the probability that it is less than –1 (as a function of p/N) appears in Table 4.3. For example, to achieve a retrieval error smaller than 1%, we have to store less than $0.185N$ patterns. Note that this is only an upper bound. In reality the storage capacity may be lower.

Indeed, Hopfield did not analyze the network capacity formally in his original paper but reported that empirical results show that the capacity of the network is about $0.15N$, similar to the theoretical capacity we have derived. Further, more careful, theoretical analysis shows that the best achievable lower bound is about $0.138N$.

This analysis indicates a relatively low capacity. If we assume strings of length 100, the network can memorize at most 15 strings using $100 \times 100 = 10,000$ weights, whereas the data can be represented by $15 \times 100 = 1500$ bits. This means that in practice the Hopfield network is of limited use for storing patterns, but it is an interesting model of distributed memory and associative recall.

4.6.7 Application of a Hopfield Network

We now describe how a Hopfield network is used to memorize the shape of digits (Haykin, 1998). The digits are represented as patterns of size 10 $\times$ 12 as can be seen in Figure 4.21. The network contains 120 neurons, where a black pixel is represented by the value +1 and an empty pixel by

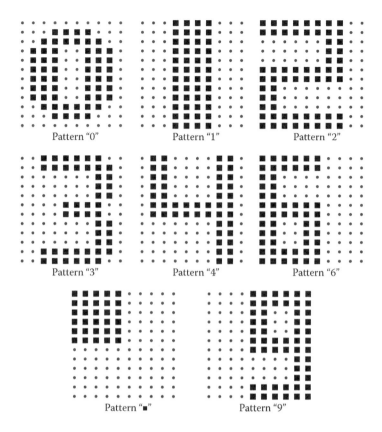

FIGURE 4.21 Example of digits memorized by the Hopfield neural network. (Adapted from Haykin, Simon, *Neural Networks: A Comprehensive Foundation*, 2nd ed. Upper Saddle River, NJ: Prentice Hall, 1998. With permission.)

the value –1. The weights are computed according to the network learning rule (4.10).

In the first phase of the experiment the memorized patterns were presented to the system, and as expected it remained in the stable configurations. In the second phase, altered patterns were presented to the system to see how it dealt with input errors. The value of each pixel in the pattern was flipped with probability 0.25. For instance, Figure 4.22 shows how the network dealt with the pattern of the digit 6, which has been altered in this fashion: it managed to converge to the memorized pattern.

Similar behavior was observed in other cases. Nonetheless, the network does fail occasionally in converging to the correct shape. For instance, in

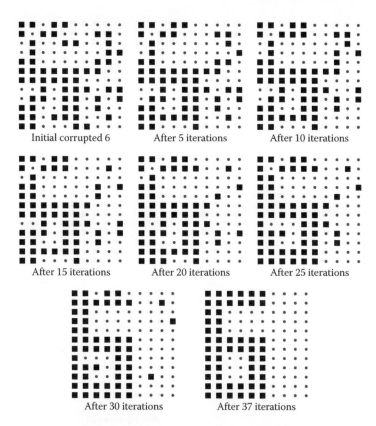

FIGURE 4.22   Example of the digit 6 where pixels were flipped with probability 0.25 that is gradually retrieved by the network. (Adapted from Haykin, Simon, *Neural Networks: A Comprehensive Foundation,* 2nd ed. Upper Saddle River, NJ: Prentice Hall, 1998. With permission.)

the example shown in Figure 4.23 the network starts with an altered representation of the digit 2 and converges erroneously to 6. A more surprising problem is the existence of stable erroneous states, that is, attractors that are not memorized patterns. These are called **spurious attractors**. This situation can be seen in Figure 4.24 where the network converged to a pattern similar to the pattern it was presented with (the digit 9 with alterations). The pattern it converged on was similar to the memorized 9 but not identical to it.

4.6.8 Further Uses of the Hopfield Network

Up to now we have dealt with the memorizing and retrieval capabilities of Hopfield networks. However, Hopfield networks have additional uses.

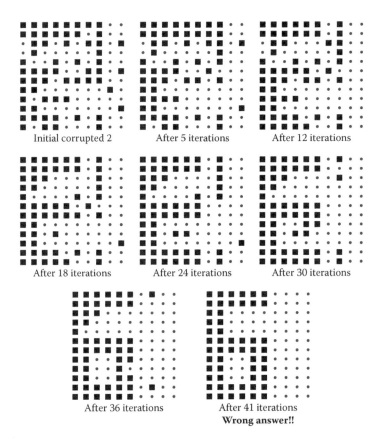

FIGURE 4.23   An example of erroneous retrieval. The network starts with a noisy version of the digit 2 (with 25% flipped pixels) and ends up retrieving the digit 6. (Adapted from Haykin, Simon, *Neural Networks: A Comprehensive Foundation*, 2nd ed. Upper Saddle River, NJ: Prentice Hall, 1998. With permission.)

Assume that we present the input patterns shown in Figure 4.25(a) to the network, and it produced the weight matrix depicted in Figure 4.25(b).

Observe the patterns to notice that the three rightmost and leftmost bits in each pattern are mirror images, while the two middle bits are independent of any other bits in the pattern. When we initialized the network with the string 0 1 0 1 1 0 0 0, after 3 update steps, computing bits from left to right, we come up with 0 0 0 1 1 0 0 0 which is the retrieval of one of the memorized patterns.

On the other hand, if we start with 1 1 0 0 0 1 1 1 we will encounter an interesting situation. After three updates steps, the string 1 1 1 0 0 1 1 1 will be generated—a string that has not been memorized but has the properties

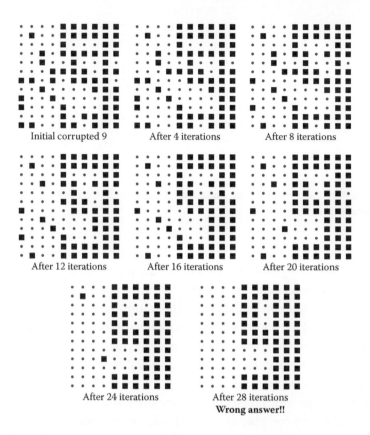

FIGURE 4.24 An example of retrieval to a spurious attractor. The network converged to the digit 9, but note that this is not exactly the same digit presented in the input data. (Adapted from Haykin, Simon, *Neural Networks: A Comprehensive Foundation*, 2nd ed. Upper Saddle River, NJ: Prentice Hall, 1998. With permission.)

we identified in the set of strings that was memorized. In other words, as the weights reflect the types of relations between the bits, the network has succeeded *in learning* the rule governing the samples. Obviously, regularity can also be learned by the feedforward networks described earlier.

Hopfield networks can also be used for optimizations. We saw in the proof of the convergence of the retrieval process that a set of weights corresponds to an energy function. This function is minimized by the process of updating the values of the neurons such that at the end of the process neurons connected by an edge with a positive weight will tend to have similar values, whereas neurons connected by an edge with a negative weight will tend to have opposite values. One can use a Hopfield network in the opposite

		0	1	0	1	0	0	1	0

(a)

	1	2	3	4	5	6	7	8
1	0	0	2	-2	0	0	0	8
2	0	0	0	2	-8	-2	8	0
3	2	0	0	-4	2	8	0	2
4	-2	2	-4	0	-2	-4	2	-2
5	0	-8	2	-2	0	2	-8	0
6	0	-2	8	-4	2	0	0	2
7	0	8	0	2	-8	0	0	0
8	8	0	2	-2	0	2	0	0

(b)

The input patterns (a):

```
0 1 0 1 0 0 1 0
1 1 0 0 0 0 1 1
1 0 1 0 1 1 0 1
0 0 0 1 1 0 0 0
0 0 0 0 1 0 0 0
0 1 1 0 0 1 1 0
0 0 1 0 1 1 0 0
0 1 0 1 0 0 1 0
```

FIGURE 4.25   An example of input patterns (a) and weight matrix (b) for a Hopfield network.

fashion: start with an energy function to be minimized, and build a network around it with weights derived from the function (a nontrivial task). If we perform a process similar to the retrieval process on this network, we will get values that are good solutions to the minimization problem. Hopfield and Tank (1986) implemented this idea for the **traveling salesman problem** (TSP). In this problem the input is a map of cities with known distances between them. A traveling salesman has to find the shortest route allowing him to visit all the cities and each city exactly once. This problem is known to be an NP-complete problem, and therefore it is commonly believed that no algorithm can compute an optimal solution in a reasonable amount of time (i.e., in time that is not exponential in the number of cities). Hopfield and Tank represented the solution as a matrix representing the order of visiting the different cities (see Figure 4.26). Since the representation does not ensure that the route is legal (e.g., two cities are designated as visited second in the route in Figure 4.26(a)), the energy function penalized for illegal routes as well as for routes that are long. The challenge was to derive a weight matrix such that the computation will end in the shortest legal route (e.g., Figure 4.26(b)). Hopfield and Tank reported good results for examples of 10 and 30 cities. Interested readers are referred to the paper for further detail.

4.7 UNSUPERVISED LEARNING

Up to now we have discussed different forms of learning in which the neural network is trained using example data to which the required output is known, and network learning is driven by comparing the output the network produces and the required output. Can learning be achieved without supplying the network with examples of previously classified data? Such learning can be achieved by analyzing the input and attempting to

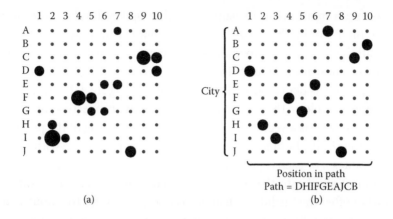

FIGURE 4.26 The representation used by Tank and Hopfield to address the TSP problem using a Hopfield network. In this binary matrix representation the first column represents which city is visited first, the second column represents the city visited second, and so forth. Note that the representation does not ensure a legal path; that is, one city can be visited multiple times, or the route can visit in two cities simultaneously. The matrix in (a) depicts such an illegal route in the beginning of the optimization process, and the matrix in (b) shows the final legal route, which happens to be the optimal route. (Adapted from Hopfield, John J. and David W. Tank, *Science* 233, no. 4764, 625–633, 1986. With permission.)

find patterns, connections, and correlations between the data points. The standard approaches to unsupervised learning are statistical in nature and attempt to single out important features of the input. We now discuss a technique based on a neural network called **self-organizing maps (SOMs)**, or **Kohonen networks**, named after Teuvo Kohonen (2001), who was the first to describe such a network.

4.7.1 Self-Organizing Maps

Self-organizing maps are used to produce a discrete low-dimensional representation of a set of input samples. The SOM network is based on **competitive learning**, where neurons compete with each other in an attempt to represent the input. The neurons are usually organized on a 2-D grid with a hexagonal structure. If each input pattern is of length N, then each neuron will hold a vector of length N. For example, if each input is a vector of three numbers representing a point in 3-D space, then each neuron will contain a vector of three numbers. The initial values of the vectors held by the neurons are random, and the network seeks the neuron that best represents each input item. When a particular neuron is selected to represent an input point, its value is further updated to be more similar to that input point. This process is iterated many times such that, when it ends, all the input points are mapped to neurons so that adjacent neurons will represent similar input points.

The process is demonstrated schematically in Figure 4.27. Assume that each input sample on the left contains values for five properties and is therefore represented as a vector of length 5. At the beginning of the process all the neurons in the hexagonal grid contain vectors of length 5 with random initial values. For each input sample we search for the neuron whose value is the most similar to that sample. "Similar" can be defined in different ways, but often the Euclidean distance is used; that is, for the input sample V_j we will search for the neuron N_i that minimizes the expression

$$\sqrt{\sum_{m=1}^{|V|}\left(V_j^m - N_i^m\right)^2}$$

where V_j^m denotes the m-th component of the vector V_j.

Clearly, we do not expect to find a good fit at the beginning of this process as the initial values of the neurons are random; nonetheless, we can pick the neuron for which the value of the previous expression is the smallest. Assume this is the neuron N_i, colored black in Figure 4.27. At this point, the value of neuron N_i is changed to be more similar to the sample V_j (see following formula). We define a neighborhood for each neuron—in

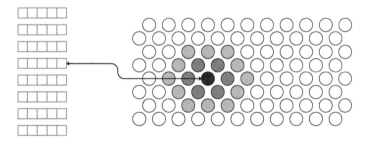

FIGURE 4.27 A schematic description of the process of creating a self-organizing map. Each input vector is mapped to a neuron of the network (which usually has the layout of a two-dimensional hexagonal grid). In an iterative procedure, the values of the target neuron as well as neurons in its neighborhood are adjusted to reflect the values of the input. Thus, the procedure achieves reduction in the dimensionality of the input—in this example, from five to two.

this example the neighborhood was chosen to be two layers deep. All neurons in the neighborhood of N_i are also updated to be more similar to V_j but to a lesser extent. So N_i is changed the most (marked in black), the neurons in the layer closest to it are changed in a weaker way (dark gray), and the neurons in the second layer are changed the least (light gray). The changes are given by the following formula:

$$N_k(t+1) = N_k(t) + \alpha(t)h_{ki}(t)[V_j - N_k(t)]$$

We see that the value of a neuron in iteration $t + 1$ is the sum of its previous value and a change term that is the product of three factors: (1) the difference between its previous value and the input sample $[V_j - N_k(t)]$; (2) the learning rate function $\alpha(t)$, which is similar to the learning constant we have encountered previously but is represented as a function since it can be changed during the computation; and (3) $h_{ki}(t)$, which determines if neuron N_k is in the neighborhood of N_i and if so how close they are to each other. Note that this factor is also time dependent and it is common to decrease the size of neighborhoods as the learning process progresses in order to refine it. Variations of this algorithm define the neighborhood and the way neighborhood sizes change with time in different ways.

After the neurons are updated, the next input point is selected, and the process is repeated. Note that the same neuron can represent more than one sample. In fact, if the number of samples is larger than the number of neurons this must be the case. This learning process, like the other learning procedures previously discussed, requires many learning epochs before it converges.

We say that the network has converged when the mapping (i.e., which sample is represented by which neuron) does not change during an entire epoch. The learning process just described leads to a **clustering** of the input data. Since neurons are arranged on a 2-D grid, we can think of the algorithm as a way to reduce the dimensionality of the data from N to 2. In contrast to some of the networks we discussed earlier, we cannot prove mathematically that this network does indeed converge, but experience shows that this process usually ends up with a stable mapping of the input samples to neurons such that similar inputs are represented by the same neuron or by adjacent neurons. Clearly, every execution of the learning algorithm will give rise to different solutions as the initial values of the neurons are chosen randomly. Each input point will be mapped to different neurons upon different choices of the initial values, but we do expect the network to be similarly organized for different initial values. As it is common to repeat this process many times, we would like to be able to identify the best mapping. For this purpose two error metrics are used: (1) the **quantization error**; and (2) the **topological error**. The first computes the average distance between each input vector and the neuron most similar to it, and it is obvious that we will prefer the maps with the smallest quantization error. The topological error computes the percentage of input samples for which the two most similar neurons are not adjacent on the grid. If the map self-organized into clusters, we expect that for most input vectors the neuron closest in value and the neuron second closest in value are adjacent on the grid, and therefore the topological error will be small. The combination of these two metrics allows us to identify the best mapping.

4.7.2 WEBSOM: Example of Using SOMs for Document Text Mining

One of the most impressive examples of using self-organizing maps is WEBSOM, a system for mining very large document collections (Lagus, Kaski, and Kohonen, 2004). The largest implementation of the system so far contains 7 million patent abstracts mapped to a network of about 1 million nodes. In this system, each document is represented by a very long binary vector where each position in the vector represents one word in a predefined vocabulary. The vector contains 1 in cells that correspond to words that appear in the document and 0 for words that do not. To reduce the huge size of such vectors, the vocabulary is made smaller by aggregating words of similar meaning. Each neuron in a 2-D SOM starts with a random vector, and the input vectors are mapped to neurons in an iterative process. When selecting the best match for a given input, the similarity between the vector representing the input and the vector represented by the neuron is calculated by counting the matches between corresponding positions in the two vectors or in mathematical terms by calculating the inner product of the two vectors. Different weights can be given to different words to reflect their relative importance to the text. Such a weighting scheme is achieved, for example, by the "inverse document frequency" measure that gives higher weights to words appearing a lot in a specific document but that are otherwise rare. The actual implementation of WEBSOM includes several shortcuts enabling the system to handle efficiently the large amount of data.

The system maps similar documents to neurons in the same region of the map. To make the map a useful text-mining tool, the system selects and automatically labels map regions. A search can start by finding documents mapped to regions whose labels match best with the search expression. Then, further relevant search results can be found by moving to documents mapped to neighboring regions even if they did not match exactly the search criteria. Figure 4.28 shows an example of such search in a WEBSOM built for 68,000 articles from the *Encyclopedia Britannica*.

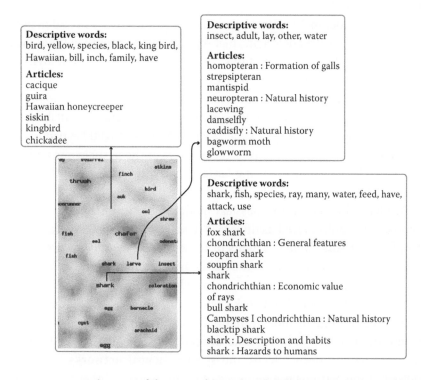

Descriptive words:
bird, yellow, species, black, king bird, Hawaiian, bill, inch, family, have

Articles:
cacique
guira
Hawaiian honeycreeper
siskin
kingbird
chickadee

Descriptive words:
insect, adult, lay, other, water

Articles:
homopteran : Formation of galls
strepsipteran
mantispid
neuropteran : Natural history
lacewing
damselfly
caddisfly : Natural history
bagworm moth
glowworm

Descriptive words:
shark, fish, species, ray, many, water, feed, have, attack, use

Articles:
fox shark
chondrichthian : General features
leopard shark
soupfin shark
shark
chondrichthian : Economic value of rays
bull shark
Cambyses I chondrichthian : Natural history
blacktip shark
shark : Description and habits
shark : Hazards to humans

FIGURE 4.28   A close-up of the map of *Encyclopedia Britannica* articles. The user has clicked a map region with the label "shark," obtaining a view of a section of the map with articles on, for example, sharks, various species of fish, and eel (middle and left); insects and larvae (lower right corner); and various species of birds (upper right corner). Searches performed on the map confirm that whales and dolphins can be found nearby (not shown). A topic of interest is thus displayed in a context of related topics. The three inserts depict the contents of three map regions, that is, the titles of articles found in the region. By clicking a title, the user can access the article. The "descriptive words" contain a concise description of the contents of each map region. (Adapted from Lagus, Krista, Samuel Kaski, and Teuvo Kohonen, *Information Sciences* 163, no. 1–3, 135–156, 2004. With permission.)

4.8 SUMMARY

In this chapter we have concentrated on the pioneering examples of neural networks developed more than two decades ago. Since then, neural networks have become practical tools in several areas including image processing and medical decisions. Probably the most popular area of application is finance, where neural networks are used to evaluate loan applications, to forecast foreign currency exchange rates, and to predict stock market behavior. One remarkable recent application of neural network is 20Q, a computerized version of the popular children game of guessing what the opponent is thinking about in 20 yes/no questions. The system uses a neural network structured as a matrix of weights that represents the strength of the association between objects and questions. This weight matrix is used to dynamically choose the next question based on previous answers, and the weights are updated when the system guesses the correct answers, reinforcing the weights involved in the successful computation. The game is available in an online version (http://www.20q.net/) with about 10,000,000 synaptic connections that keep learning from the answers of the participants and also as a small handheld device using far fewer weights and without the ability to learn. Try it and you will be amazed by the performance of the system.

We have focused on the classical models of neural networks: the perceptron, the multilayered feedforward–backpropagation network, and the Hopfield network. Many variants of these models as well as several new models have been suggested. Among them are **recurrent networks**, where the network includes not only forward edges but also backward connections; **stochastic neural networks**, where the output of the neurons is not deterministic and includes random noise; **dynamic neural networks**, where the network exhibits time-dependent behavior such as transient phenomena and delay effects; and **spiking neural networks**, where the output of the neurons is a sequence of pulses (spikes) rather than an output of a constant level. Spiking neural networks are inspired by biological neurons that can represent information not only by the level of the output but also by the rate in which the pulses are emitted. It is too early to tell if any of these models or other emerging models will achieve the prominence of the models we discussed in this chapter.

In this chapter we discussed learning and memory. We demonstrated how similar principles to those used by the central nervous system can be implemented by simple computational systems. Clearly, the biological systems are different and much more complex than the artificial systems we dealt with, but we hope we have succeeded in highlighting the similar principles, first and foremost of which is the importance of high connectivity between the basic building blocks of the systems. This connectivity is at the core of the capabilities of the human brain and is the common basis of all the artificial neural networks we have described.

Will future artificial networks be as capable as the human brain? The jury is still out on this issue. Many researchers believe that there is a vast difference between the two kinds of systems. They believe that we are incapable of building and controlling systems with a similar number of components and level of connectivity as exists in the human brain. According to this view not only is the limitation technological; we also simply do not understand well enough the actual data processing mechanisms that occur in the brain and the principles governing these mechanisms to enable us to simulate them in an artificial system. On the other hand, some researchers believe that most of the difference between the biological and artificial processing capabilities is merely quantitative, and when we consider the rate of growth of computational systems (as measured by, e.g., the number of components, their speed and complexity, memory size) we see that the rate of technological advancement is so fast that it will soon catch up with biological systems. In the bibliography you will find a reference to one of the leaders of this school of thought, **Ray Kurzweil** (2005).

4.9 FURTHER READING

Chittka, Lars and Jeremy Niven. 2009. Are bigger brains better? *Current Biology* 19, no. 21, R995–R1008.

Haykin, Simon. 1998. *Neural Networks: A Comprehensive Foundation,* 2nd ed. Upper Saddle River, NJ: Prentice Hall.

Hopfield, John J. 1982. Neural networks and physical systems with emergent collective computational abilities. *Proceedings of the National Academy of Sciences* 79, no. 8, 2554.

Hopfield, John J. and David W. Tank. 1986. Computing with neural circuits: A model. *Science* 233, no. 4764, 625–633.

Kohonen, Teuvo. 2001. *Self-Organizing Maps,* 3rd extended ed. New York: Springer.

Kohonen, Teuvo, Samuel Kaski, Krista Lagus, Jarkko Salojarvi, Vesa Paatero, and Antti Saarela. 2000. Self organization of a massive document collection. *IEEE Transactions on Neural Networks* 11, no. 3, 574–585. Available at: http://websom.hut.fi/.

Kurzweil, Ray. 2005. *The Singularity Is Near: When Humans Transcend Biology.* New York: Viking.

Lagus, Krista, Samuel Kaski, and Teuvo Kohonen. 2004. Mining massive document collections by the WEBSOM method. *Information Sciences* 163, no. 1–3, 135–156.

LeCun, Yann, L.D. Jackel, B. Boser, J.S. Denker, H.P. Graf, I. Guyon, et al. 1989. Handwritten digit recognition: Applications of neural network chips and automatic learning. *IEEE Communications Magazine* 27, no. 11, 41–46.

Marr, David. 1982. *Vision: A Computational Investigation into the Human Representation and Processing of Visual Information.* New York: W. H. Freeman and Co.

Minsky, Marvin and Seymour Papert. 1969. *Perceptrons.* Cambridge, MA: MIT Press.

Rumelhart, David E., Geoffrey E. Hinton, and Ronald J. Williams. 1986. Learning internal representations by error propagation. In D. E. Rumelhart, J. L. McClelland, and the PDP Research Group (Eds.), *Parallel Distributed Processing. Explorations in the Microstructure of Cognition. Volume 1: Foundations,* 318–362. Cambridge, MA: MIT Press.

Segev, I. 1998. Sound grounds for computing dendrites. *Nature* 393, no. 6682, 207–208.

Sejnowski, Terrence J. and Charles R. Rosenberg. 1987. Parallel networks that learn to pronounce English text. *Complex Systems* 1, 145–168.

Tesauro, Gerald and Terrence J. Sejnowski. 1989. A parallel network that learns to play backgammon. *Artificial Intelligence* 39, no. 3, 357–390.

4.10 EXERCISES

4.10.1 Single-Layer Perceptrons

1. Study the OR function learned by a simple perceptron with three inputs x_0, x_1, x_2 and threshold of 0. Fill out the values of the weights in Table 4.4 using the learning rule. Let the value x_0 always be –1. The patterns are presented to the perceptron in the order in which they appear in the table. The learning rate is $\alpha = 0.5$. Repeat as needed until the weights converge.

TABLE 4.4

X_0	X_1	X_2	W_0	W_1	W_2	Output	Desired Output
–1	0	0	1.3	0.4	–0.2		
–1	0	1					
–1	1	0					
–1	1	1					
–1	0	0					
–1	0	1					
–1	1	0					
–1	1	1					
–1	0	0					

2. Design a perceptron for the following problem: the inputs to the network are strings of six bits (i.e., there are six input neurons), and the perceptron has six output neurons that count the number of bits, which are 1 in the input strings as follows:

The output is 000000 if the input contains no 1's.

The output is 100000 if the input contains a single 1 (its position is unimportant).

The output is 110000 if the input contains two 1's.

...

...

The output is 111111 if the input is all 1's.

3. Can you design a perceptron for Exercise 2 if the output is represented by the position of a single 1 in the output neurons? For instance:

100000 represents an input string containing a single 1.

010000 represents an input string containing two 1's.

000010 represents an input string containing five 1's.

Justify your answer.

4.10.2 Multilayer Networks

4. Design a network that deals with images of the format shown in Figure 4.29. Every image contains N cells that are either white or black. The network has to determine whether the image contains more than N_0 contiguous black regions. The images are circular (the first and last cells are considered to be adjacent). For instance, in Figure 4.29, the top image contains two black regions, and the other images contain one black region each. Suggest a topology for the network, and determine its weights. Your solution can disregard the special case where all cells are black.

FIGURE 4.29

5. Design a multilayered network that implements the XNOR function. (XNOR is Not XOR; that is, it returns 0 when XOR returns 1 and 1 when XOR returns 0).

6. Given the network in Figure 4.30, determine the weights $W_1,...,W_9$ so that the network computes correctly the XOR function on binary inputs A and B. (Note that each hidden neuron has a control input with the fixed value –1.) To simplify matters, you may assume that all the neurons compute using a threshold function (as in perceptrons) rather than a sigmoid function.

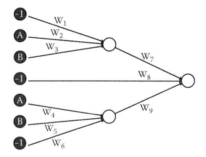

FIGURE 4.30

7. The network in Figure 4.31 is to be used to compute the XOR function. The learning rate is $\alpha = 0.5$, and the activation function is a sigmoid. Perform a full epoch of weight updates for the inputs (0,1) and (1,1) where the initial weights are $w_1 = -0.1$, $w_2 = 0$, $w_3 = 0.1$, $w_4 = 0.2$, $w_5 = 0.1$, $w_6 = -0.2$, $w_7 = 0.1$, $w_8 = -0.3$, and $w_9 = 0$.

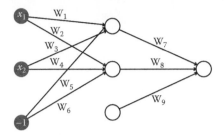

FIGURE 4.31

8. In the NetTalk network:

    a. What are the advantages and disadvantages of a larger window?

    b. How can the optimal window size be determined?

9. In the handwriting recognition example:

    a. Why was such a large training set needed?

    b. Calculate the number of connections between the neurons in the first hidden layer and the neurons in the input layer (consider the threshold level as input) and the number of weights needed to specify these connections.

10. Consider a multilayered network with seven input neurons, seven output neurons, and n neurons in a single hidden layer. For simplicity, assume that the neurons in the hidden layer function according to step function and not the sigmoid function. The network is designed to map unary representations of the digits 0 to 7 to identical outputs (the bottleneck method). In this representation the input patterns 0 is represented as 0000000, the input 1 as 10000000, 2 as 01000000, and 7 as 00000010.

    a. What is the minimal number of neurons in the hidden layer?

    b. Design a network to solve this problem.

4.10.3 Hopfield Networks

11. Consider the Hopfield network shown in Figure 4.32 where the threshold values of the neurons are 0.

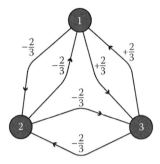

FIGURE 4.32

a. Display the weights as a matrix T where the cell in position (i,j) holds $T_{i,j}$.

b. Show that the weights in T adhere to the requirements on weights in a Hopfield network.

c. The network contains three neurons, allowing for eight possible patterns. The stored states are $(1,-1,1)$ and $(-1,1,-1)$. Compute the behavior of the network on these patterns, and show that they are stable.

d. For each vector obtained by changing the value of a single neuron in the stored patterns (e.g., changing $(1,-1,1)$ to $(-1,-1,1)$, $(1,1,1)$ or $(1,-1,-1)$, determine the vector to which the network will converge. Does the network deal well with errors?

12. A Hopfield network containing five neurons has to store the following patterns:

$$U_1 = (1, 1, 1, 1, 1)$$
$$U_2 = (1, -1, -1, 1, -1)$$
$$U_3 = (-1, 1, -1, 1, 1)$$

a. Determine the 5×5 weight matrix for the network.

b. Show that the stored patterns are stable states of the network.

c. Check the network's behavior when presented with a noisy version of U_1 where the second element is –1.

d. Show that the following patterns are also stored in the network. What is the relationship between these patterns and the original patterns?

$$U_1 = \left(-1, -1, -1, -1, -1\right)$$

$$U_2 = \left(-1, 1, 1, -1, 1\right)$$

$$U_3 = \left(1, -1, 1, -1, -1\right)$$

13. Design and implement a Hopfield network that memorizes digits and retrieves them. Every digit will be represented as a 10 × 10 matrix containing 0's and 1's. The network will memorize the digits and will retrieve a digit when presented with its image with a few flipped bits. For instance, in Figure 4.33 the network will retrieve the digit 3 on the left when presented with the image on the right, which has several altered bits. Explore a few aspects of this network:

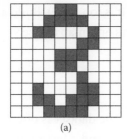 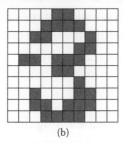

(a)   (b)

FIGURE 4.33

a. Determine how many different digits the network can memorize. A digit is considered memorized if the network can retrieve it correctly in 90% of the cases in which 10% of the bits are flipped.

b. Explore the trade-off between the number of memorized digits and the number of altered bits. Plot a three-dimensional graph where X is the digit axis, Y is the percentage of errors axis, and the Z axis is the success percentage.

c. Repeat the tests where every digit is memorized using several similar but not identical input patterns (create the patterns first). Does this improve the learning?

d. Repeat the experiments by representing the digits using 1 and –1 (rather than 1 and 0). Does this improve the success rate or the convergence rate? Why?

e. The network can fail in one of two ways: it can converge to a digit that was learned but that is not the correct digit, or it can converge to a state that was not part of the input samples (a spurious attractor). Count the number of errors from each type, and print out a few samples of states that are not part of the input samples. Try to classify those patterns.

4.10.4 Self-Organizing Maps

14. Download a red/green/blue (RGB) color table. In it each color is coded using three numbers that represent the strengths of its red, green, and blue components. For instance, navy blue is represented as (65,105,225) and pink as (255,192,203). Such tables can be found on many Web sites. Make sure your table contains about 100 colors (choose a sampling of the table if it is too large). Build a self-organizing map that will classify the colors on a map of size 9 × 9 and display the results graphically.

4.10.5 Summary

15. In supervised learning the size of the dataset is often limited. Thus, dividing the data into a training set and a test set might render the training set too small. Suggest a way to handle such situations.

16. Due to the complexity of designing multilayer neural networks (e.g., the need to determine the number of layers, the number of neurons at each level, weight sharing), it has been suggested that genetic algorithms can be used to find good network architectures. Discuss how to implement this idea (consider the components required to characterize the genetic algorithm, such as data kept in chromosomes, as well as the components required to characterize the network, such as the learning process).

We presented several potential applications of neural networks (of all types). For Exercises 17–22 discuss how to represent the input for each problem in an appropriate way and which network should be used to solve the problem. Note that these problems can be approached in different ways and that no single correct solution or method can be found that guarantees finding a successful network without using trial and error.

17. Optimization problems:

    a. The map coloring problem: given N countries and K colors, color the countries such that no two adjacent countries (countries sharing a border) have the same color. Countries may remain uncolored, and the goal is to minimize the number of uncolored countries.

    b. The N queens problem: place N queens (or less) on an $N \times N$ chessboard such that no queen can attack another queen. The goal is to minimize the number of unplaced queens.

18. Design a network that suggests reasonable past forms for English verbs when given the present tense of the verbs. Keep in mind that many English verbs have irregular past tenses (e.g., run–ran, speak–spoke, draw–drew, die–died). Consider data representation, network design, and expected deficiencies in the network's learning and generalization capabilities.

19. Image processing:

    a. Face recognition: the goal is for the network to identify a person from a photograph. Assume the photograph is a headshot.

    b. Image reconstruction: the network is presented with images that have some missing regions (e.g., parts were torn off) and some regions that are out of focus. The network is to reconstruct the originals from these partial images.

20. Suggest a design for a network to forecast failures in a large engine in a power plant. The network uses sensors that report the temperature, number of rotations per minute, fuel flow, and vibrations as well as a microphone that captures the engine's sound. Under normal circumstances the engine operates continuously (and the network design has

to take this into account). For training purposes, however, the engine may be stopped, and started; failures can be induced, etc.

21. A bank uses an expert system to determine which clients are credit-worthy. The system is based on parameters such as the clients' age and gender, their balance sheets, number of operations per month, and the type of requested credit (i.e., whether it is for small, medium, or large loans). The system has been in use for a few years, and the bank has data on its performance (i.e., the cases where loans were approved and turned out to be bad loans.) Will a neural network be able to improve the loan approval process? Suggest an appropriate architecture.

22. Identifying unusual credit card usage. Design a network for a single customer (one credit card). Note that under normal circumstances there are no unusual usage patterns of the sort the network needs to identify that can be used for training the network.

4.11 ANSWERS TO SELECTED EXERCISES

2. A perceptron where each one of the six inputs is connected to all of the six output neurons and the threshold of the first output neuron is 1, the threshold of the second output neuron is 2, and so forth accomplishes the given task.

3. This is impossible to do with a simple perceptron. Look at a smaller example of two inputs and two outputs. The behavior we require for the first output bit is to be 1 if and only if one of the two input bits is 1; this is the XOR function that cannot be calculated with a simple perceptron.

4. The black cells will be represented by the value 1 and the white cells by 0. The network has a single hidden layer as shown in Figure 4.34. For the hidden layer a simple step function with a threshold of 0 is used. Thus, a node in the hidden layer will fire only when it recognizes a switch from a white cell to a black cell. Note that because of the circular nature of the problem we have a node that connects the first cell and the last cell. The output neuron will have a threshold of N_0, so it will output 1 only if there are at least N_0 contiguous black regions.

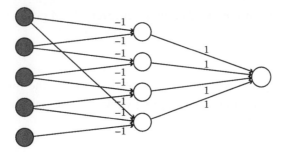

FIGURE 4.34

8.

a. The network has more data the larger the window and therefore can pronounce the letter better. If, for example, there was
no window around the letter, the network would be unable to
distinguish between the k's in "kill" and "know." On the other
hand, a larger window means that the network is more complex
and has to learn a larger number of weights. That will cause the
training phase to be longer, require a larger training set, and may
result in an impractical convergence rate.

b. The decision has to be based on the network's performance: can
it achieve a low enough percentage of errors to be useful?

9.

a. The variety of different shapes is very large, and the number of
neurons and weighs grows proportionally. Moreover, the large
variability in the ways the same digit can be written necessitates
the use of a large training set to decrease the danger of overfitting, where the network would be well trained on the training set
but incapable of generalizing.

b. The hidden layer contains 768 neurons (8 × 8 × 12), which have
19,968 connections. The number of weights is only 1068 (768
threshold values + 12 × 25 weights). This means that the network has to learn ~5% of the weights it would have had to learn
if weight sharing was not employed. The second hidden layer is
similar to the first. It contains 12 groups of 4 × 4 neurons that
are "in charge" of a region of size 5 × 5 in the first hidden layer.
The neurons get input from corresponding regions in 8 of the 12

groups in the first hidden layer (every 4 × 4 group is "in charge" of a different combination of eight groups).

10

a. Since the network actually needs to count the numbers 0 to 7, three neurons are enough to capture the data using a binary representation.

b. A possible design for the network is shown in Figure 4.35. The input nodes are connected to the hidden layer nodes in a way that represents the binary coding; for example, every second input bit is connected to the upper hidden node that represents the parity (rightmost) bit in the binary representation. The weight of all edges from the input to the hidden layer is 1 (to prevent overloading the figure not all of these weights are shown). The threshold values of all the hidden layer neurons are 1. The hidden layer neurons are fully connected to the output neurons (again, to prevent overloading the figure, not all of these edges are shown). The reconstruction is achieved by setting the appropriate values for the weights on the edges and by the threshold of the output units. The weights are set according to the binary representation of the number; for example, the three edges going into the fourth neuron, which represents the number 4 with the binary representation 100 should have weights of (1,–1,–1). The threshold value for each neuron is the number of 1's in the corresponding binary representation, which is 1 in the case of 100.

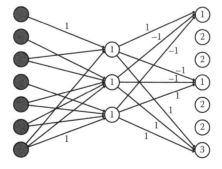

FIGURE 4.35

11.

a.

$$T = \frac{1}{3} \begin{bmatrix} 0 & -2 & +2 \\ -2 & 0 & -2 \\ +2 & -2 & 0 \end{bmatrix}$$

b. We have to test for two conditions: (1) that the weights T_{ii} are equal to 0; and (2) that the weights satisfy Hebb's rule (i.e., formula (4.10)). Verify that T satisfies both of these conditions.

c. We will demonstrate this using matrix multiplication notation.

$$T\mathbf{v} = \frac{1}{3} \begin{bmatrix} 0 & -2 & +2 \\ -2 & 0 & -2 \\ +2 & -2 & 0 \end{bmatrix} \begin{bmatrix} +1 \\ -1 \\ +1 \end{bmatrix} = \frac{1}{3} \begin{bmatrix} +4 \\ -4 \\ +4 \end{bmatrix}$$

Applying the sign() function yields:

$$\text{sign}(T\mathbf{v}) = \begin{bmatrix} +1 \\ -1 \\ +1 \end{bmatrix} = \mathbf{v}$$

which means that **v** is stable. The calculation for the other vector is similar.

15. A possible solution is to use all the available data for training, leaving out only a single example to be used for testing. To evaluate the performance of the network in this way, one needs to repeat the procedure many times, each time leaving out a different example for testing, and then evaluate the performance over all trials. This is known as the "jackknife" method. Note, however, that the data points have to be independent and different from each other in order for this technique to work. If many data points are duplicated, then when one copy is presented as the test case, the other copy is included in the training test and would taint the result of the test

CHAPTER 5

Molecular Computation

THIS CHAPTER DEALS WITH solving computational problems using chemical and biological processes. Problems to be solved using molecular computation are usually presented as a collection of molecules that are mixed together and undergo a series of biological processes. These processes produce a new collection of molecules that represent the solutions to the computational problem. This approach to solving computational problems is interesting in several ways:

- **Practical motivation**: Molecular computation may allow us to build computational devices from biological molecules and possibly to build general-purpose "biological computers." As we will discuss in this chapter, such devices may have several advantages over classical computers for certain applications. An interesting example might be in medicine where such devices may be able to make autonomous real-time decisions inside the patient's body.

- **Theoretical motivation**: The observation that biological molecules can carry out computations should help us realize that many biological processes involved in information processing and biological control should be viewed as computational processes. The computational perspective can help us gain better insight into these biological processes. Furthermore, this viewpoint opens the way to employ the tools and methodologies of computer science to analyze biological processes.

- **Parallelism**: Molecular computations are inherently parallel processes, as they involve a large number of molecules that collide and interact with each other. Thus, molecular computational models are potentially very powerful and may be able to solve hard computational problems.

- **Analog computation**: Molecular computation is analog computation rather than digital computation. Historically, the first computing machines were mechanical and were used for astronomical computations. They made use of gears, cams, levers, drums, and other mechanical components to perform complex computations such as computing integrals. The heyday of the mechanical computing machines was during World War II when electromechanical machines called BOMBE were used to crack the Enigma code used by the Germans. This era came to an end when electronic computers using binary logic took over and replaced the mechanical machines. The term analog computation, however, encompasses a wider range of computational processes based on making use of other physical phenomena like hydraulics, optics, and biological phenomena to solve computational problems. In other words, analog computers use the physical behavior of a system to solve computational problems. Consider, for example, the following. When two boards connected by rods are dipped in soap water, bubbles with a minimal surface area will be created between the rods, because closed physical systems will reach an equilibrium state of minimal energy, and thus, in this example, the system will minimize the surface of the bubbles. We could harness this phenomenon to solve the following computational problem. Given a set of vertices, interconnect them by edges such that the total length of the edges (the sum of the lengths of all the edges) is minimized. You may add vertices and edges to the graph to reduce the total length. (The added vertices are called Steiner vertices, and the problem is known as the **Steiner tree problem** and is an NP-complete problem.) The soap bubbles self-organize by adhering to the laws of physics to find the Steiner tree defined by the location of the rods between the boards (Figure 5.1).

  To date, analog computations are hardly considered by computer scientists, but they do raise very interesting algorithmic, complexity, and computability questions such as how to measure the complexity of an analog system, what can and cannot be computed using an

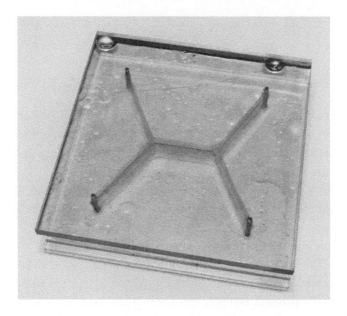

FIGURE 5.1   Soap bubbles self-organize to create the optimal Steiner tree on a four-vertex graph. (Picture courtesy of Scott Grandison.)

analog computer, and to what extent the analog and digital computational models are equivalent. We believe that the new computational models described in this chapter, along with quantum computing (out of the scope of this book; a computational model based on qubits, which can be in a state of one, zero, or quantum superposition of zero and one) will raise interest in analog computation.

5.1 BIOLOGICAL BACKGROUND

In this section we describe a number of laboratory techniques used in the molecular computations discussed later in this chapter.

5.1.1 PCR: Polymerase Chain Reaction

Polymerase chain reaction (PCR) is a major player in biological lab work and has a central role in the molecular algorithms described in this chapter. PCR is used to amplify a DNA segment of interest multiple times. Kary Mullis developed this technique in the 1980s and was awarded the 1993 Nobel prize in chemistry for this accomplishment.

PCR makes use of the enzyme **DNA polymerase** (Figure 5.2), which completes a single strand of DNA, which is used as a template for a

FIGURE 5.2 The DNA polymerase reaction.

double-stranded DNA molecule. To perform its function, the enzyme has to encounter single-stranded DNA that terminates with a double-stranded segment at one end. The DNA polymerase elongates the double strand to the entire molecule. When DNA is heated to a high temperature (around 95°C), the strands separate from each other, whereas they attach to each other at lower temperatures in a process called **hybridization** or **annealing**.

The combination of strand complementation, separation, and annealing allows the PCR process to turn into a chain reaction. A DNA sequence to be amplified is selected, and short sequences of DNA called **primers** are prepared. The primers complement the beginnings of each selected DNA strand and must be roughly 20 bases long for the PCR to succeed. The PCR process starts when DNA, the primers, the bases used to create new strands, and the DNA polymerase are mixed together in a test tube. The complete process is described in Table 5.1.

A **thermal cycler**, or a **PCR machine**, is used to automatically heat and cool the solution in the test tubes. Note that the solution should contain enough "raw material" so that the process can go through all its

TABLE 5.1 The PCR Process

1. The DNA strands are separated by heating them to around 95°C.	
2. The mixture is cooled to around 55°C and the primers attach themselves to the beginnings of the corresponding strands (base pairing).	
3. The mixture is warmed to 72°C which is optimal for the functioning of the Polymerase. The enzyme complements the bases on each strand, thereby doubling the number of DNA molecules.	
4. The whole process is repeated from Step 1 as often as needed. The number of available DNA molecules is doubled at each iteration.	

phases. PCR allows researchers to start from a very small amount of DNA (obtained, e.g., from fossils or found at a crime scene) and to amplify it quickly. For example, if we want to know whether a sample of DNA contains a particular DNA sequence, we can prepare unique primers that surround the sequence and start a PCR reaction. If the sequence is present in the sample it will be amplified; otherwise, the two primers will not attach, and the reaction will not take place.

5.1.2 Gel Electrophoresis

Gel electrophoresis is a technique for separating molecules such as DNA and proteins using an electric field applied to a gel. Different molecules move differently in the gel according to their size and electrical charge, so smaller and higher-charged molecules will be more affected by the electric field and therefore will move faster. Since DNA molecules have a similar charge to mass ratio, the main difference between molecules is their size, which determines their ability to migrate through the pores of the gel. Thus, running DNA molecules on a gel sorts the molecules according to their size and allows researchers to determine the size of new DNA molecules by comparing them with DNA molecules of known sizes (which form a **DNA ladder** when run on the gel).

5.1.3 Restriction Enzymes

Restriction enzymes cut double-stranded DNA molecules. They operate by binding to the DNA at a specific **restriction site** (a short sequence usually four to eight bases long) and incising the DNA at the site or close to it. Some enzymes perform a **blunt** incision, and others leave behind **sticky ends** that allow the ends to join other DNA molecules (Figure 5.3). The first restriction enzyme was discovered by Werner Arber, Dan Nathans, and Hamilton Smith, who were awarded the Nobel prize in medicine in 1978. It is believed that these enzymes evolved in bacteria to protect them against viruses with double-stranded DNA. Restriction enzymes have turned into essential molecular biology tools,

FIGURE 5.3 Restriction enzymes. (a) Blunt edge. (b) Sticky end.

FIGURE 5.4   The ligation process. (a) Before. (b) After ligation the double-strand structure is complete.

since they allow cutting DNA for many purposes such as introducing new sequences into existing DNA molecules. Hundreds of restriction enzymes are currently available commercially.

5.1.4 Ligation

Ligase enzymes can repair breaks in one of the strands of a DNA molecule (provided that the molecule is held together by its double-strand structure). This is done by bonding (ligating) adjacent nucleotides, thereby recovering the original structure of the molecule (Figure 5.4). Ligase enzymes are used by cells both in DNA repair and in DNA replication. In molecular biology, ligase and restriction enzymes are often used in concert to introduce new sequences into DNA molecules.

5.2 COMPUTATION USING DNA

5.2.1 Hamiltonian Paths

Molecular computation seemed almost like science fiction until **Leonard Adleman** published his paper "Molecular Computation of Solutions to Combinatorial Problems" in 1994. In the paper, Adleman presented an implementation of a molecular process to solve the classical computer science problem of finding **Hamiltonian paths** in a directed graph:

> Let $G = (V,E)$ be a directed graph and v_{in} and v_{out} be two of its vertices. A **Hamiltonian path** is a path starting at v_{in} and terminating at v_{out} that goes through every vertex exactly once. Given G, v_{in}, and v_{out}, determine whether there exists a Hamiltonian path in G.

The Hamiltonian path problem is known to be NP-complete; therefore, solving it using DNA was an exciting development.

We first present an abstract *nondeterministic* algorithm to solve the problem, which we will then use as an outline for the molecular algorithm.

The algorithm generates random paths, which are tested to determine if they are Hamiltonian. The algorithm follows these steps:

1. Generate a large set of random paths in the graph, where a path is a set of one or more edges where the starting vertex of an edge has to match the ending vertex of the previous edge in the path.

2. Discard all paths that do not start at v_{in} and do not terminate at v_{out} as they cannot be a solution to the given Hamiltonian path problem.

3. Discard all paths whose length (the number of vertices they traverse) is not equal to the number of vertices in the graph, as they cannot be a solution to the given Hamiltonian path problem. Note that this stage may retain paths that do not solve the problem as they may visit some vertices more than once and never visit other vertices.

4. Discard all paths that do not visit every vertex as these paths cannot be solutions to the Hamiltonian path problem. Note that we do not need to check separately if the remaining paths visit some nodes more than once, as this follows from Steps 3 and 4.

5. If the resulting set is nonempty, return "yes." Otherwise, return "no."

This algorithm has two main stages: (1) generate a set of candidates; and (2) sieve out all the candidates that do not solve the problem. This is called a **generate and test** algorithm, and the molecular algorithm operates in a similar way.

The main trick in implementing the algorithm using molecules is the way vertices and edges are represented. We represent the graph as a collection of single-stranded DNA molecules, where the vertices and edges are represented as follows (Figure 5.5):

$$
\begin{array}{ll}
O_0 & \text{TATAGGGGTAGCGCTTTTGC} \\
O_2 & \text{TATCGGATCGGTATATCCGA} \\
O_3 & \text{GCTATTCGAGCTTAAAGCTA} \\
\overline{O_0} & \text{ATATCCCCATCGCGAAAACG}
\end{array}
$$

$$
\begin{array}{ll}
O_{2\cdot3} & \text{GTATATCCGAGCTATTCGAG} \\
O_{0\cdot3} & \text{TATAGGGGTAGCGCTTTTGCGCTATTCGAG}
\end{array}
$$

FIGURE 5.5   Molecular representation of vertices and edges in a graph.

- **Vertices**: Every vertex is represented by an arbitrary sequence of 20 bases (we could have chosen another arbitrary number of bases). The sequence associated with each vertex is arbitrarily chosen, but we will see that these sequences may not be identical, similar, or complementary to sequences associated with other vertices. There are 4^{20} possible such representations, which is assumed to be much larger than the number of vertices in the graph. The molecular representation of vertex i will be denoted as O_i. The complementary sequence, where A complements T and C complements G, will be denoted as $\bar{O}_i$.

- **Edges**: Each edge is also represented by a sequence of 20 bases. The edge between vertices i,j is composed of the last 10 bases of i's representation followed by the first 10 bases of j's representation. We will see how this allows us to connect the edges to create paths. The molecular representation of the edge $i \to j$ will be denoted as $O_{i \to j}$.

    The edges starting at v_{in} or ending at v_{out} have a slightly different representation. They use the complete representation of these special vertices, so these edges will be represented by sequences of length 30. An edge from vertex v_{in} to vertex j will be represented by the entire representation of v_{in} followed by the first 10 bases of j. Similarly, edge from vertex i to vertex v_{out} will start with the last 10 bases of the representation of i followed by the entire representation of v_{out}.

The significant property of this representation is that it enables the connection of adjacent edges into a sequence that represents a path. This is achieved by putting together, in a solution, molecules representing edges and molecules representing the complements of the vertices. So if, for example, we introduce the representation of the edges $O_{2 \to 3}$ and $O_{3 \to 4}$ and the complement of O_3 denoted as $\bar{O}_3$, $\bar{O}_3$ will combine with the second half of the edge $O_{2 \to 3}$ and with the first half of the edge $O_{3 \to 4}$ to create a *double-stranded sequence*, as can be seen in Figure 5.6.

When such double-stranded sequences are created in the presence of the enzyme ligase, the two separate molecules $O_{2 \to 3}$ and $O_{3 \to 4}$ will be ligated to form a single molecule representing the path $2 \to 3 \to 4$. In this manner all possible paths in the graph are generated from the molecules representing the edges of the graph; then, they can be sieved such that only the Hamiltonian paths (if they exist) remain. Another example, representing the path $0 \to 3 \to 4$, appears in Figure 5.7. Note that this process can create sequences representing paths that are not legal Hamiltonian paths. The molecular

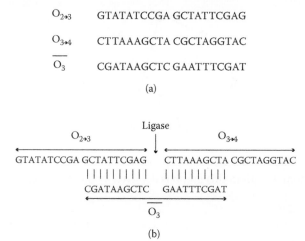

FIGURE 5.6   Building the path 2 → 3 → 4. (a): The participating molecules: two edges, and the complement of a vertex. (b): The segment of the path that can be created by these molecules.

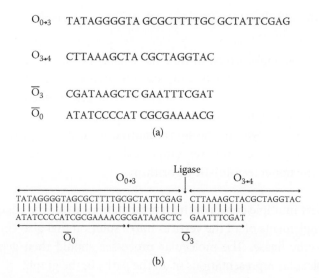

FIGURE 5.7   Building the path 0 → 3 → 4. (a): The participating molecules: two edges, and two complements of vertices. (b): The segment of the path that can be created by these molecules. Note that, because of the longer structure of the edge $O_{0 \to 3}$, the beginning of the path has a blunt shape that cannot be extended.

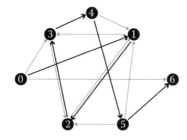

FIGURE 5.8   The graph used by Adleman to compute a Hamiltonian path. A possible Hamiltonian path in this graph is depicted by the darker arrows.

$O_0 \rightarrow O_1$		$O_1 \rightarrow O_2$	$O_2 \rightarrow O_3$	$O_3 \rightarrow O_4$	$O_4 \rightarrow O_5$		$O_5 \rightarrow O_6$
$\bar{O}_0$		$\bar{O}_1$	$\bar{O}_2$	$\bar{O}_3$	$\bar{O}_4$	$\bar{O}_5$	$\bar{O}_6$

FIGURE 5.9   A legal Hamiltonian path in Adleman's graph.

$O_0 \rightarrow O_2$		$O_2 \rightarrow O_5$		$O_5 \rightarrow O_6$
$\bar{O}_0$		$\bar{O}_2$	$\bar{O}_5$	$\bar{O}_6$

FIGURE 5.10   An illegal Hamiltonian path in Adleman's graph.

algorithm was implemented in Adleman's paper on the seven-vertex graph in Figure 5.8, where vertex 0 is the starting vertex v_{in}, and vertex 6 is $v_{out.}$ Figure 5.9 depicts a legal Hamiltonian path in the graph, whereas Figure 5.10 shows a path through the graph that is not a legal Hamiltonian path.

Let us now go over all the steps required for executing the molecular algorithm for the seven-vertex graph, using the same steps we used to describe the nondeterministic algorithm:

1. Insert multiple copies of the representations of all the edges and the complements of all the vertices into a solution that also contains the enzyme ligase. The molecular processes should then generate the molecular representations of all the paths in the graph.

2. At this point, a process that creates many copies of the paths starting and ending at the appropriate vertices is applied to the solution. This is achieved by doing a PCR with primers that are the molecular representations of v_{in} (O_0) and the complement of v_{out} ($\bar{O}_6$).

3. The DNA molecules are separated based on their length, and only the molecules of length 140 ($20n$, where n is the number of vertices in the graph) are kept. As each edge is represented by a molecule of length 20 and a Hamiltonian path has to traverse each vertex exactly once, a Hamiltonian path will contain $n - 1$ edges. But as the edge starting at v_{in} and the edge ending at v_{out} are of length 30, the total length of the representation of the Hamiltonian path is $20n$. The molecules that were kept are amplified again so that enough copies exist to continue with the algorithm.

4. The paths that visit every vertex in the graph are selected. This is achieved by first selecting from the solution all the molecules that contain the sequence O_1. The selection of such molecules is achieved by first heating the solution to separate the DNA strands and then applying magnetic beads attached to sequences that are complementary to those being searched for and thus anneal with them. Next, the selected molecules are further processed to select only the ones containing O_2 and so on, up to O_5. The molecules we are left with at the end of this process represent paths that visit each vertex.

5. PCR is applied again to the solution (using the same primers) to make sure that even a small number of appropriate DNA molecules will be detectable.

6. Gel electrophoresis is performed to test whether any DNA molecules of the appropriate length are found. If such molecules exist in the solution, this proves that a Hamiltonian path exists in the graph, and the algorithm outputs *yes*; otherwise the output is *no*.

Figure 5.11, taken from Adleman (1994), shows the gel used to read out the composition of the DNA molecules found. The gel shows the final step of the algorithm (Step 6). To produce the image in Figure 5.11, a number of PCR processes were applied to the molecules, each of which contained O_0 as one of its primers; the complements of O_1 through O_6 were used consecutively as the other primer. The resulting DNA was then run on a gel. Each lane in the figure shows the result of the amplification of a specific subsequence of the Hamiltonian path. This process allows us to see the formation of the molecules for all partial paths and to prove that the resulting Hamiltonian path is $0 \rightarrow 1 \rightarrow 2 \rightarrow 3 \rightarrow 4 \rightarrow 5 \rightarrow 6$. Alternatively, it is possible to sequence the resulting molecule and actually read out the sequence of the vertices.

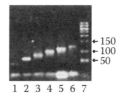

FIGURE 5.11 The experiment proving the existence of a Hamiltonian path: Graduated PCR of the final product of the experiment revealing the Hamiltonian path (lanes 1 through 6); the molecular weight marker is in lane 7. (From Adleman, Leonard M., *Science* 266, no. 5187, 1021–1024, 1994. With permission.)

In Adleman's experiment the implementation of the molecular algorithm for a seven-vertex graph took about a week of lab work. Some of the steps can be automated, and the time to implement such an algorithm may be shortened to a matter of hours; indeed, in subsequent works Adleman presented apparatuses that can do this. Nonetheless, some of the processes are time-consuming; for example, the PCR steps cannot be significantly shortened. Clearly, for a seven-vertex graph the computation time is much shorter on a regular computer. However, computer scientists are interested in analyzing how the time to execute an algorithm changes as a function of the problem's size. The molecular algorithm was run on a small graph, and we would like to know how its speed would vary as the size of the graph is increased. This allows us to neglect "fixed costs" and focus on the execution speed as a function of the input size (the number of vertices in the graph). The complexity of the algorithm is analyzed by counting the change in the number of basic operations the algorithm goes through as a function of the input size (usually denoted by n). Note that counting basic operations is better than measuring actual running time, as it is not dependent on the speed of the machines used to execute the algorithm.

The number of lab steps needed for implementing the molecular algorithm grows linearly with the size of the input. That means that if the number of vertices is doubled, the number of lab steps will also double (ignoring fixed costs that do not change as a function of n). This can be seen by looking at all the steps taken by the molecular algorithm and analyzing how they depend on the size of the graph. For most of the steps it seems that the amount of time is constant regardless of the size of the graph. For example, the hybridization performed in Step 1 would take roughly the same time whether 7 or 7000 molecules are involved. The only step that is directly dependent on the size of the graph is Step 4 where the number of selection operations is the same as the number of vertices in the graph, so obviously the number of selection operations will double when the number of vertices is doubled. If at every step of the algorithm the increase in the number of operations is equal (or is a *constant* multiple) to the increase in the number of vertices, the total increase is linear, and the algorithm is called a *linear* algorithm.

All the conventional Hamiltonian path algorithms require a number of operations that grows very fast—in fact, exponentially—as a function of the size of the graph. This is very different from the behavior of the molecular algorithm, which is more and more cost-effective as graphs gets larger.

Counting the number of steps needed to execute the algorithm in the lab allows us to estimate the efficiency of the algorithm but hides the number of concurrent operations caused by the collisions of the molecules in the test tube. As already stated, parallelism is the secret behind the power of the molecular algorithm, and to appreciate its strength it is important to try to determine how many molecular operations are performed at each stage. This will allow us to estimate the number of basic operations and to facilitate the comparison between the performance of the molecular computation and regular digital computations.

Let us estimate the number of ligation operations performed in Step 1 of the molecular algorithm to evaluate the number of operations performed per time unit, which is the common metric of the speed of a computational device. This approach provides a way of quantifying the advantage a molecular algorithm has relative to a sequential computation using a digital computer. Adleman (1994) estimated that the number of ligation operations performed in Step 1 was on the order of 10^{14} and that it would be relatively easy to increase this number to at least 10^{20}. Dividing this into the length of time it takes to perform Step 1 (about an hour) shows that the molecular computation operations are about as fast as basic operations performed on a supercomputer.

It is common in computer science to evaluate algorithms by looking at the time and memory needed to execute them (as a function of input size). One could also look at other resources. For molecular algorithms we might want to look at the cost of resources, such as the number of molecules needed to perform the algorithm or the amount of energy consumed by the process, again as a function of the graph size.

The number of distinct molecule *types* needed for implementing the algorithm (i.e., representing the vertices and edges), grows only linearly with the graph size. However, this is not the case for the total number of molecules. It is important to realize that the algorithm has to generate all possible paths in the graph to make sure that each one is tested for being Hamiltonian. Since the number of possible paths of length N in the graph is roughly d^N (where d is the average degree of nodes in the graph), the amount of DNA (i.e., the total number of molecules) needed will grow exponentially with the number of vertices. Assume a graph with 80 vertices and an average degree of 10. We will then need more than 10^{80} DNA molecules. This number is larger than the estimated total number of atoms in the universe, which obviously indicates that this and similar algorithms are not practical for large problems. Thus, these types of molecular algorithms do not really offer us a magic bullet to shatter the limits of solving large exponential problem. Nevertheless, these algorithms may be useful in addressing medium-size problems that are not amenable to conventional computation.

Another disadvantage of the molecular algorithm, compared with digital computation, is that it is susceptible to errors. As we will see, errors can arise in various stages of the algorithm and can adversely affect the success of the algorithm. This is particularly dangerous in Step 4 of the molecular algorithm. If a single valid path fails to be selected and is removed due to an error, the algorithm may return a *false negative*, even though a Hamiltonian path does exist. That is the reason for repeating the PCR at each step—amplifying the number copies representing each solution reduces the probability that all the copies will be lost due to an error. The danger of a *false positive* is not as problematic, as the molecules that gave rise to the positive answer are validated at the final step to verify that they indeed represent a Hamiltonian path in the graph. If it turns out that the molecule does not represent a Hamiltonian path, other molecules from the final batch can be tested.

A major source of errors during the molecular computation is due to the potential creation of wrong double-stranded molecules. For instance, a sequence s might hybridize with a sequence that is similar but not identical to its complement $\bar{s}$ because of a partial match. Another possibility is that single strands that are complementary on a subsequence will create a double-stranded molecule even though they are not complementary on the full length of the sequence: a sequence xy can create a partially double-stranded molecule with the strand $\bar{y}z$ by pairing the sub-sequences y and $\bar{y}$. Another case is when two regions in the single strand happen to complement each other, thereby losing the linear structure and creating loops in the molecule.

The more similar is the representation of a vertex to the complement of another vertex, the larger is the probability of creating erroneous double-stranded sequences. To lessen this danger, it is better to choose molecular representations that minimize the probability of creating such errors, that is, choosing representations that are different enough from each other. This can be achieved by using longer molecular representations and selecting the sequences so that the number of complementing bases between any two representations will be minimal.

Note that the hardware of digital computers is also prone to various errors, which might occur because of electromagnetic induction between components of the electronic circuits or because of electrical noise in the vicinity of the circuit. The errors may affect the computation of the logical circuits, but the main danger lies in the fact that they may change the content of stored data bits. Various techniques are applied to minimize these risks, and digital computers are usually considered so trustworthy that programmers and users tend not to consider the possibility of hardware errors. Nonetheless, when building computers that have to be extremely reliable (e.g., aircraft control systems), system architects do address the possibility of hardware errors—both at the physical design and at the logical design (e.g., algorithmic) levels.

To conclude, let us compare the molecular algorithm with a digital computer:

1. **Speed**: Personal computers currently operate at speeds of Gflops (10^9 floating-point operations per second). The fastest supercomputers are close to achieving speeds of Petaflops (10^{15} floating-point operations per second). Adleman (1994) estimated that one can achieve a similar number of 10^{15} ligation operations per second. In other words, the number of basic operations per second (which are not particularly expensive or complex) of the molecular algorithm is similar to that of a digital supercomputer.

2. **Memory** (space efficiency): A single bit of information is stored in a molecular representation with a volume of about 1 cubic nanometer. Modern disks can store 10^{11} bits in a cubic centimeter that translate to one bit in 10^{10} cubic nanometers; that is, the molecular representation offers a dramatic improvement.

3. **Energy**: One joule of energy is enough for 2×10^{19} ligation operations, which is 10^{10} more efficient energy-wise than a supercomputer.

(A **joule** is a physical unit of energy equaling roughly the amount of energy required to lift a 1 kilogram object 10 centimeters off the Earth's surface.)

4. **Flexibility**: The biggest problem is that the use of molecular algorithms requires a radically different design of the algorithm for each problem (in terms of, e.g., representation, logical operations, and laboratory procedures) and that in each case the algorithm needs to be created from very basic molecular operations. Contrast this with solving problems using digital computers where modern high-level programming languages and software design tools enable solving computational tasks in fairly straightforward ways (at least most of the time).

5.2.2 Solving SAT

We now present a molecular algorithm for solving another computationally hard problem called the **satisfiability** problem, or **SAT** for short. This is also an NP-complete problem, like the Hamiltonian path problem. The solution described here was presented by **Richard Lipton** in 1995 and uses Adleman's (1994) technique for constructing all possible paths in graph as part of its construction (Lipton, 1995).

The particular case of SAT known as **3SAT** has an illustrious computer science history, as it was the original problem proven directly to be NP-complete. All the subsequent NP-completeness proofs are based on reducing 3SAT to the problem or reducing another problem that was reduced directly or indirectly from 3SAT. The molecular solution we present here is not as "elegant" as Adleman's Hamiltonian path solution (even though it is based on it) and exposes a brute-force characteristic that is much more subtle in Adleman's work.

Definition of the SAT Problem

Let $U = \{u_1, u_2, \ldots u_n\}$ be a set of logical variables. A logical variable's value can be either "true" or "false." An assignment is a function t that determines for every element in U a value, either "true" or "false," denoted from now on by 1 and 0, respectively. If u is a variable from the set U, then u and u' (meaning "not u," also denoted by $\neg u$) are called literals. The literal u has the value "true" under the assignment if and only if the variable u was assigned the value "true." Similarly, the literal u' has the value "true" if and only if the variable u was assigned the value "false."

The SAT problem is defined as follows:

Given a logical formula of the form $C = C_1 \wedge C_2 \wedge ... \wedge C_m$, where every clause C_i is of the form $v_1 \vee v_2 \vee ... \vee v_h$, and each v_i is a literal, determine whether there exists an assignment for the variables in C for which the logical formula C is true (i.e., its value is 1). We say that such an assignment *satisfies* the formula C.

Examples

- Let $C = (p)$. This formula contains one clause and one variable, and the assignment $\{p = 1\}$ satisfies it.

- Let $C = (p \vee q)$. This formula contains one clause and two variables, and any assignment where either p or q assumes the value 1, such as $\{p = 1, q = 0\}$, satisfies it.

- Let $C = (p) \wedge (q)$. This formula contains two clauses and two variables. In this case the assignment has to be $\{p = 1, q = 1\}$ for both clauses to be satisfied.

- Let $C = (p \vee q) \wedge (p' \vee q')$. Here we have two clauses: $(p \vee q)$ and $(p' \vee q')$. To satisfy the formula, both clauses have to be satisfied simultaneously. The two assignments $\{p = 1, q = 0\}$ and $\{p = 0, q = 1\}$ both satisfy the formula.

- Let $C = (p) \wedge (p')$. It is easy to see that an assignment cannot exist that satisfies C because it is impossible for both p and p' to be true at the same time.

- Let $C = (x \vee y \vee z) \wedge (x \vee y \vee z') \wedge (x' \vee y' \vee w) \wedge (x \vee z' \vee w')$. For C to be true, we have to assign values to the variables x, y, w, and z such that for each clause there exists at least one literal with the value 1. Here $\{x = 1, y = 1, w = 1, z = 0)$ is one such an assignment. This formula is an instance of a 3SAT problem, where each clause contains exactly three literals.

To solve SAT we have to represent all possible assignments for the variables in the formula. To do so, DNA sequences representing all the binary strings of a given length are prepared. For instance, if the formula contains three logical variables x, y, and z, all the eight possible Boolean assignments have to be evaluated.

In general, 2^n representations of sequences have to be generated for n variables. To do this efficiently we make use of the power of

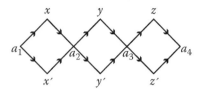

FIGURE 5.12   The graph generating all binary sequences of length three.

molecular computation using similar techniques to those used for solving the Hamiltonian path problem. We will build a graph in which each path from the initial node a_1 to the last node a_n represents a binary string. The graph for $n = 3$ is shown in Figure 5.12.

An example of a path in the graph is $a_1 x a_2 y' a_3 z a_4$. This represents the string 101: at each node a_i one can choose between the upper and lower edges. The upper edge corresponds to choosing 1 as the assignment of one variable, whereas the lower nodes represent the bit 0. Each path represents an assignment of the set of variables. So the path $a_1 x a_2 y' a_3 z a_4$ represents the assignments $\{x = 1, y = 0, z = 1\}$. As in Adleman's (1994) algorithm, we choose random DNA sequences for the vertices a_1, a_2, a_3, and a_4 and for the vertices representing the variables x, y, and z and their negations x', y', and z'. Note that the representations of x, y, and z and their negations x', y', and z' are not complementary DNA sequences (e.g., the representation of x does not complement the representation of x'). Using these representations for the vertices, the edges are then built as described in Adleman's algorithm. All the possible paths in the graph will be generated in a test tube containing all of these molecules together with the enzyme ligase, similar to the way the paths were generated in the Hamiltonian path problem. Each path represents one distinct truth assignment for the n variables in the formula.

Note that in such a graph Adleman's technique will create all the paths representing exactly all the 2^n combinations of the variables' values. After they are generated it is possible to select all sequences with the value 1 in the i-th bit or all sequences without 1 in that bit, that is, sequences with the value 0 in the i-th bit. This can be achieved by using magnetic beads attached to the complements of the sequences of interest. The general strategy is to prepare all possible assignments as previously described and then to select the sequences that satisfy the formula (clause by clause) and to discard sequences that do not satisfy the clauses. If (and only if) any sequences remain, we know that a satisfying assignment exists for the formula.

We start with a test tube marked T_0 containing all possible assignments to the variables. The computation proceeds to create a set of $T_1,T_2,...,T_m$ of test tubes (where m is the number of clauses in the formula) such that T_i contains only the sequences that satisfy the clauses $C_1,C_2,...,C_i$.

Algorithm Structure

- Prepare a test tube T_0 containing all possible variable assignments.

- For all $1 \le i \le m$ perform the following:

  - Generate the test tube T_i from the molecules in the test tube T_{i-1} by selecting the sequences representing assignments that satisfy the clause C_i (the exact way of doing it is left as an exercise; see Exercise 5.10).

    Recall that the sequences in tube T_{i-1} contain only sequences satisfying all the clauses $C_1,...,C_{i-1}$; thus, this process will iteratively eliminate sequences that do not match the clauses of the logical formula.

- Test the tube T_m to see if it contains any DNA (there are many simple ways for doing this). If so, the algorithm returns "*yes.*" If not, the formula is not satisfiable, and the algorithm returns "*no.*"

Performance Analysis

If the formula contains m clauses and is made up of a total of l literals, then the number of operations the molecular SAT algorithm performs is linear in l. If the number of literals per clause is fixed (as in the 3SAT case), then the performance of the algorithm is linear in m. We leave the proof of these assertions as exercises for the reader (see Exercises 5.11 and 5.12).

5.2.3 DNA Tiling

In the 1960s **Hao Wang** suggested using tiling as a computational mechanism. This model seems simple but turns out to be very challenging. In the tiling model there is a finite number of types of tiles and an infinite number of tiles of each type. Each tile has four labeled sides (e.g., the sides may be colored), and tiles may be laid next to each other only if any two touching sides are labeled the same. The tiles may not be rotated. Given such a set of tiles, the computational task is to determine if it can be used to tile the entire plane (Figure 5.13). In some cases the answer is trivial. For instance, if all tiles are of the same kind and the opposite sides share

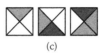

(a)   (b)   (c)

FIGURE 5.13   Tile samples. (a): Single tile that can be used for tiling the plane. (b): A pair of tiles that cannot tile the plane. (c): Three tiles for which the problem is difficult.

a color (Figure 5.13(a)), then it is possible to use them to tile the plane. In other cases (Figure 5.13(b)) it is obvious that they cannot be used to tile the plane. But in many cases (Figure 5.13(c)) the answer is far from obvious. In fact this problem is equivalent to the halting problem we discussed in Chapter 2. In other words, in general it is impossible to find an algorithm that will decide if the tiling problem has a solution in finite time. Moreover, it is possible to build the equivalent of a Turing machine using such tiles; therefore, the tiling model can be used to solve any problem solvable using a Turing machine; that is, it is computationally universal.

Erik Winfree (Winfree et al., 1998) suggested that DNA molecules can be used to represent "tiles." Winfree demonstrated the power of the tiling model by using it to build a binary counter (Barish, Rothemund, and Winfree, 2005) as demonstrated in Figure 5.14.

Winfree used seven types of tiles, depicted at the top of Figure 5.14. The tile labeled S is the starting tile. The two shaded tiles are used to build the frame in which the computation occurs, and the remaining four tiles are used for the computation. (This time, in contrast to Figure 5.13, we match

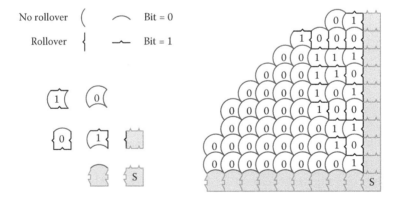

FIGURE 5.14   A binary counter using tiles. (Adapted from Barish, Robert D., Paul W. K. Rothemund, and Erik Winfree, *Nano Lett* 5, no. 12, 2586–2592, 2005. With permission.)

shapes instead of colors.) The computing tiles represent binary values— two represent the value 0, and the other two represent the value 1.

The computation starts by laying down the starting tile S (this is the only time it is used). Subsequently, the framing tiles necessarily have to be laid out as depicted in Figure 5.14. Using this frame, the local rules specifying how tiles match will lead the computing tiles to self-organize in a way that creates a binary counter where the first row represents the value 1, the second row the value 10, followed by 11, and so forth. The computation can be understood by studying Table 5.2: a tile has the same value as that of the tile directly underneath it if there is no carry (denoted by a smooth face) and has the opposite value if there is a carry (the face has a protrusion). So whenever a carry causes the value to change from a 1 to a 0, the corresponding tile will have a protrusion, thereby passing the carry bit to the tile on its left.

As mentioned already, it is theoretically possible to use tiling to perform any computation; thus, if we can implement tiles by using DNA molecules, we will be able to use DNA to perform any computation. However, achieving this in practice presents many technical challenges. DNA molecules are usually viewed as having binary recognition capabilities; that is, the two strands can recognize each other. If we can construct DNA molecules such that four strands can mutually recognize each other, then we can directly implement the tile model. By coding the molecules correctly we can expect

TABLE 5.2   The Matching Rules of Tiles

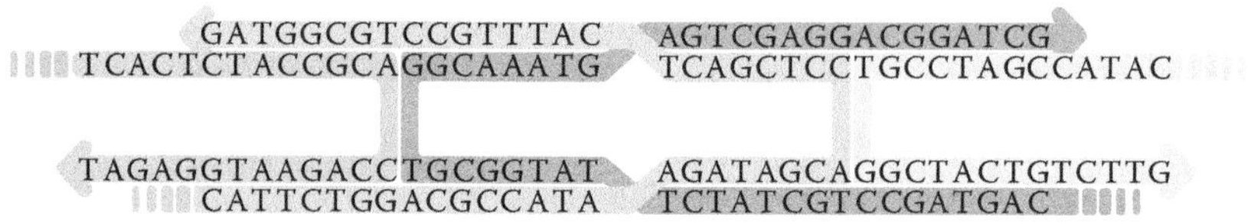

```
GATGGCGTCCGTTTAC    AGTCGAGGACGGATCG
TCACTCTACCGCAGGCAAATG    TCAGCTCCTGCCTAGCCATAC


TAGAGGTAAGACCTGCGGTAT    AGATAGCAGGCTACTGTCTTG
CATTCTGGACGCCATA    TCTATCGTCCGATGAC
```

FIGURE 5.15   A schematic DNA structure for creating "tiles" that can join other tiles. (Adapted from Barish, Robert D., Paul W. K. Rothemund, and Erik Winfree, *Nano Lett* 5, no. 12, 2586–2592, 2005. With permission.)

that they will self-organize in a way that will solve the tiling problem. For instance, in the previous example we would like the DNA molecules to recognize each other if and only if two recognition surfaces (on the bottom and on the right) match.

Winfree et al. (1998) created complex DNA structures that implement four-way recognition (Figure 5.15). Each structure contains four DNA molecules that are partly single stranded and partly double stranded. Using these molecules, Winfree designed a relatively simple tiling representing a two-dimensional periodic template and implemented it in the lab. Later he also implemented the binary counter as well as other complex computations. Nonetheless, the necessary self-assembly processes have a low yield (i.e., not all molecules self-organize when they should) and a high error rate (i.e., molecules are matched even if they should not match).

5.2.4 DNA Computing—Summary

We have discussed Adleman's (1994) technique for solving computational problems using DNA. This technique is useful when attempting to solve NP-complete combinatorial optimization problems such as the Hamiltonian path problem. We also saw how to use this technique to solve another NP-complete problem, SAT.

In both cases the generate-and-test algorithms have two main phases:

1. "Possible solutions" are created randomly.

2. The solution set is refined so that only true solutions to the problems are kept.

The first phase uses the fact that single-stranded DNA molecules create double-stranded sequences according to the base-pairing rules. So, for instance, in Adleman's Hamiltonian path algorithm the edges combine

TABLE 5.3   The Basic Molecular Operations

Operation	Logical Meaning
Extraction	Extract all molecules containing a given sequence.
Length	Separate the molecules according to their length (using gel electrophoresis).
Pour	Combine the content of two test tubes without changing the molecules.
Amplify	Create many copies of molecules or segments using PCR.
Anneal (base pairing)	Create double-stranded molecules from single-stranded molecules using base pairing.

to create paths. These molecules, which represent possible solutions to the problem, are generated randomly and in parallel as a result of collisions between molecules in the test tube.

The second phase is based on a sequence of molecular operations. Each operation is performed simultaneously and in parallel on all possible solutions in the test tube. This is reminiscent of parallel computing technique called **single instruction, multiple data (SIMD)**.

It is useful to catalog the set of basic molecular operations used to build this kind of algorithm (Boneh et al., 1996) and to understand their logical purpose (see Table 5.3). So far this toolbox of molecular operations has not allowed us to create a practical high-level language to facilitate easy construction of molecular algorithms for solving many diverse computational problems.

Winfree presented another way of computing using DNA; he created tiles that can be matched in ways that implement complex computations. As tiling is a universal computational model, this approach should allow us to solve any solvable computational problem using molecular computation. In practice, the translation of computational problems into tiling problems is nontrivial, and this approach is fraught with technical difficulties.

5.3 ENZYMATIC COMPUTATION

In Adleman (1994) we saw how to represent a computational problem as a collection of DNA molecules and to compute using a sequence of lab operations. In Winfree et al. (1998) we saw how to use DNA to implement a molecular computation that proceeds in the test tube without the need for external manipulation. The approach we describe next, developed by **Ehud Shapiro's** group (Benenson et al., 2001) is similar in this respect.

The fact that a molecular computation is autonomous is of major importance as it may allow us to build "molecular computers" that will operate independently as components in biological and chemical processes not requiring external manipulations. An example of such a process may be a biological device for drug release inside the body. Moreover, this approach allows us to regard autonomic biological processes as computational processes, a point of view that has theoretical importance when considering life as a computation.

5.3.1 Finite Automata

This section presents a molecular computing device that implements a **finite automaton**. A finite automaton is a model of a system that reacts to a sequence of inputs. At any given time period the machine is in one of a finite set of states, and the transitions from state to state depend on the next input being read.

The automaton's set of states is denoted as Q, and the states are q_i, where i is between 0 and n. The input tape contains an input word that is a sequence of symbols from a finite alphabet denoted by Σ. The automaton's "read head" reads this sequence of symbols, and the automaton changes its state accordingly. At any given point during the computation the reading head is located on a particular input symbol (*current input*). After the symbol is read, a basic computation step composed of two actions is performed:

1. The automaton proceeds to the next state (which may be the same as the current state).

2. The input tape advances so that the read head is placed on the next input symbol.

The automaton repeats this basic computation step until the tape has been read entirely. The final state, which the automaton reaches after reading the complete tape, is defined as the result of the automaton to the input word. The states of the automaton are of two kinds: (1) *accepting;* and (2) *nonaccepting.* If the last state the automaton reaches after reading the complete word is an accepting state, we say that the automaton has accepted the input word. Otherwise, the automaton rejects the input. We denote the set of accepting states by F. The set of all words accepted by the automaton is called the *formal language accepted by the automaton.*

To ensure a unique computation outcome, one of the states is defined to be the *initial state* and is denoted as q_0.

The formal definition of a deterministic finite automaton is as follows: Let $\Sigma = \{a_1, a_2, \ldots, a_n\}$ be a finite alphabet. A deterministic finite automaton $A = (Q, q_0, \delta, F)$ over Σ is characterized by the following four parameters:

1. Q is a finite set of states.

2. $q_0 \in Q$ is the initial state of the automaton.

3. $\delta: Q \times \Sigma \to Q$ is a deterministic function that defines the transition between the automaton's states.

4. $F \subseteq Q$ is a finite nonempty set of terminal (or accepting) automaton states.

The automaton is an abstract state machine that changes its state as dictated by the function δ when reading an input string with symbols in Σ.

If the automaton is in state $q_j \in Q$ and it reads the symbol $a_i \in \Sigma$, it will transition to the state $\delta(q_j, a_i)$. If $a_{i_1} a_{i_2} \ldots a_{i_k}$ is a sequence of length k of symbols over Σ then $\delta(q_j, a_{i_1} a_{i_2} \ldots a_{i_k}) = \delta(\delta(q_j, a_{i_1}), a_{i_2} \ldots a_{i_k})$.

A language L is defined by an automaton A over an alphabet Σ as the set of all strings over Σ for which the automaton transitions from the initial state q_0 to some terminal state $q \in F$, that is, $L(A) = \{x | \delta(q_0, x) \in F\}$.

It is customary to distinguish between two kinds of finite automata. In a **deterministic finite automaton** there is a single transition for any combination of the current state and the current input symbol. In contrast, for a **nondeterministic finite automaton** there is a set of possible next states for any combination of the current state and the current input symbol. The set may be empty, in which case the automaton cannot continue reading the input. A nondeterministic automaton accepts all words for which there exists any path from the initial state to an accepting state. It is also common to allow transitions that do not advance the reading head on the input tape—these are called *ε-transitions*. Surprisingly, it turns out that the set of languages defined by deterministic finite automata is identical to the set of languages defined by nondeterministic finite automata.

The following examples show that finite automata can be represented in several ways:

- In the automaton described in Figure 5.16, q_0 is both the initial and the only accepting state. The arrows indicate that upon reading either a or b the automaton transits from q_0 to q_1 and back upon reading the next symbol. The automaton accepts all even-length words and no odd-length words.

- A finite automaton that accepts strings with an even number of the symbol b (Figure 5.17).

- A formal description of a finite automaton:

$$\Sigma = \{a,b\}$$

$$Q = \{q_0, q_1, q_2\}$$

$$F = \{q_1\}$$

$$\delta(q_0, a) = q_0$$

$$\delta(q_0, b) = q_1$$

$$\delta(q_1, a) = \delta(q_1, b) = \delta(q_2, a) = \delta(q_2, b) = q_2$$

The language defined by this description is

$$L(A) = \{a^k b \mid k \geq 0\}$$

So the automaton accepts all words of the form *b, ab, aab, aaab,* and so forth.

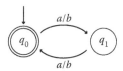

FIGURE 5.16   An automaton that accepts all strings over $\Sigma = \{a,b\}$ of an even length.

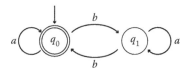

FIGURE 5.17   An automaton that accepts all strings over $\Sigma = \{a,b\}$, where the symbol b appears an even number of times.

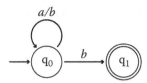

a/b

q_0 b q_1

FIGURE 5.18   A nondeterministic automaton that accepts all strings over $\Sigma =$ $\{a,b\}$ that end with the symbol b.

- The automaton in Figure 5.18 is nondeterministic as there are two possible transitions for q_0 upon reading the symbol b, and there is no possible transition from state q_1, regardless of the next symbol. If this state is visited before reaching the end of the input word, the automaton gets stuck and the word is rejected. When the automaton is in state q_0 and reads the symbol b, it has to "guess" whether to make the transition to q_0 or to q_1. If the automaton guesses that this is the last input symbol it makes the transition to the accepting state q_1; otherwise, it stays in state q_0. For every word in the language (i.e., every word ending with b), there is a computation that accepts it (i.e., where the guess is correct), although alternative computations may exist that do not terminate in an accepting state. Conversely, for words that are not in the language, there is no possible computation that will end in an accepting state. Note that if the automaton reaches state q_1 before reading the complete word, it halts in this state (as there are no transitions out of it), and the word is not accepted (to accept a word it has to be read in its entirety).

- One can also view the computation performed by a finite automaton as an execution of a set of derivation or rewrite rules on strings as follows. Assume that we are given the string q_0bab, which contains the initial state, and we are given the following rules:

$$q_0a \rightarrow q_0$$

$$q_0b \rightarrow q_1$$

$$q_1a \rightarrow q_1$$

$$q_1b \rightarrow q_0$$

where each rule encodes a given state and the left-most symbol of the string and specifies what to do in that combination. For example, the first rule states that if the string starts with q_0a it should be replaced with the symbol q_0. For the string q_0bab we start by applying the second rule to obtain the string q_1ab. Now we can apply the third rule to derive the string q_1b and finally the string q_0 by using the fourth rule. At this point there are no rules we can apply, so q_0 is the final product of the chain of derivations. Note that this is exactly the same automaton we saw in Figure 5.17. If we define q_0 to be a terminal state, the automaton will accept all strings containing an even number of occurrences of the symbol b. The string in the example contains an even numbers of b's, and indeed the computation ended in the state q_0. We will soon see how molecular computations can implement finite automata described by derivation rules.

5.3.2 Enzymatic Implementation of Finite Automata

The analogy between automata with a read head that traverse an input tape and the biological process of a DNA sequence being "read" during operations such as transcription is the reason many researchers attempt to connect the two models. Here we describe one of these attempts done by Shapiro's group—implementing finite automata using molecular techniques. The automata described here are simple two-state automata operating on an alphabet containing two symbols.

Shapiro's computation is performed on DNA molecules sequentially cleaved by enzymes until the final result of the computation is obtained. The restriction enzyme allowing for this automatic cutting without the need for external manipulation is called **FokI**.

The FokI enzyme is a **restriction enzyme** (see Section 5.1.3); that is, its function is to cleave a DNA molecule. It operates by recognizing a particular sequence of nucleotides (called a **restriction site**) and by cutting the DNA at a location a few bases from the recognition site. Let us look at the DNA molecule in Figure 5.19. The enzyme recognizes the sequence GGATG and cleaves that strand 9 bases away and the other strand 13 bases away. As a result of this particular type of cut, the resulting double-stranded DNA molecule has a single-strand end that is four bases long (i.e., a **sticky end**). The result is depicted in Figure 5.20.

The finite automaton is implemented by representing each transition rule as a DNA molecule. These molecules are called **transition molecules** and are similar to the derivation rules previously described. The

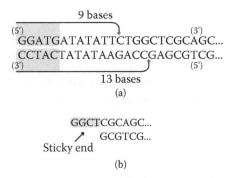

(5')   (3')
GGATGNNNNNNNNN
CCTACNNNNNNNNNNNNNN
(3')   (5')

FIGURE 5.19   FokI recognition site and cutting pattern (*N* can be any nucleotide).

9 bases

(5')   (3')
GGATGATATATTCTGGCTCGCAGC...
CCTACTATATAAGACCGAGCGTCG...
(3')   (5')

13 bases

(a)

GGCTCGCAGC...

↗   GCGTCG...

Sticky end

(b)

FIGURE 5.20   The result of a FokI restriction. (a) Binding of FokI. (b) The result-ing molecule.

input string is represented as a DNA molecule chosen such that the FokI enzyme will cleave it and expose the appropriate sticky ends. A transition molecule attaches to the sticky end in accordance with both the current symbol in the input string and the current state of the automaton, and is ligated to the molecule. This representation is the basis of the automaton's molecular implementation.

The computational process follows these steps:

1. Representing the input string as a double-stranded DNA molecule containing the FokI recognition site.

2. The FokI enzyme cleaves the molecule and exposes the sticky end.

3. A transition molecule representing one of the transition rules attaches to the sticky end and determines the next state of the automaton. It is important to make sure that only one molecule can attach, thereby implementing the required transition rule. As we discuss later, if more than one molecule is allowed to attach to the sticky end, a nondeterministic automaton is in effect implemented.

4. Repeat from Step 2.

Steps 2 through 4 continue to take place as long as the FokI enzyme can find a restriction site and as long as the appropriate transition molecules exist. If the transition molecules are depleted, the automaton halts and does not reach an accepting state. The molecular representation of the automaton described in Figure 5.17 is depicted in Figure 5.21. Following the molecular computation in this example is left as an exercise (Exercise 5.17).

The key idea of the molecular implementation of the finite automaton is finding a molecular representation that combines the automaton state and the next input symbol. This representation allows one DNA molecule both to represent the automaton's input and to implement its memory during the computation. First, we choose representations for each symbol in the alphabet, namely, for *a* and *b*, as well as for a special *terminator* symbol that will signal the end of the input string. Each representation is six bases long. A possible representation is shown in Table 5.4.

The current state of the automaton is stored in the DNA molecule by having each of the automaton states represented by exposing a different

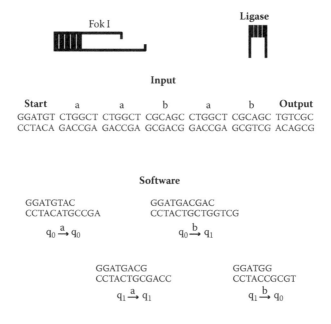

FIGURE 5.21 Molecular computation. The figure shows the molecular representation of the input string, the transition molecules, and the two enzymes needed to implement an automaton that accepts binary strings with an even number of *b*'s. (Courtesy of Ehud Shapiro.)

TABLE 5.4   Representation of the Automaton's Alphabet

A	B	**Terminator** (t)
CTGGCT	CGCAGC	TGTCGC

TABLE 5.5   Representation of Symbol–State Combinations

State	Symbol		
	A CTGGCT	**B** CGCAGC	**T** TGTCGC
q_0 $\Delta = 2$	$<q_0,a>$ CT<u>GGCT</u>	$<q_0,b>$ CG<u>CAGC</u>	$<q_0,t>$ TG<u>TCGC</u>
q_1 $\Delta = 0$	$<q_1,a>$ <u>CTGG</u>CT	$<q_1,b>$ <u>CGCA</u>GC	$<q_1,t>$ <u>TGTC</u>GC

sticky end of the representation of the next input symbol. The sticky ends depend on the representation of the symbol and the current state of the automaton. For state q_1 we use the first four bases of the symbol's representation, and for q_0 we use the last four bases of the representation. The displacement from the beginning of the symbol's representation is denoted by Δ and is in one-to-one correspondence with the state of the automaton. In Table 5.5 the sticky end is underlined for every symbol–state combination.

Recall that the state transition function determines the next state as a function of the current symbol and state: $\delta : Q \times \Sigma \rightarrow Q$. Thus, when FokI cleaves the last added transition molecule, it will expose a sticky end representing the new state of the automaton. The next transition molecule will attach itself to this sticky end as dictated by the new state and the next input symbol and so on until the input has been read in its entirety.

Each transition molecule has three components (see, e.g., the four transition molecules in Figure 5.21):

1. The FokI recognition site.

2. The region identifying the current state and symbol.

3. The *spacer* region determining the next state. This component is the key for the proper implementation of the automaton, as its length determines the automaton's next state (Table 5.6).

TABLE 5.6   Spacer Length Representation of the Automaton's State Transitions

Spacer Length	State Transition
1	$q_1 \rightarrow q_0$
3	State unchanged
5	$q_0 \rightarrow q_1$

How does the length of the spacer achieve the desired result? Recall that FokI removes nine bases after the recognition site and that each symbol is represented by six bases. These, together with the current displacement Δ, determine what will be the displacement from the beginning of the next symbol arising after the FokI cleavage. This displacement uniquely determines the identity of the next state of the automaton.

$$\Delta_{new} = 9 - (6 - \Delta_{current}) - spacer = 3 + (\Delta_{current} - spacer)$$

Plugging the result of this formula into each of the transition states produces the displacements shown in Table 5.7. The table confirms that, given the displacement of the current state, the displacement produced after the cleavage matches the new state as required. So by choosing the spacer length it is possible for the transition molecules to encode the state transitions. This construction allows us to create molecules describing all eight possible transition rules for a two-state automaton over a two-symbol alphabet (Figure 5.22). When interested in recognizing a particular language (e.g., all strings containing an even number of occurrences of the symbol b), one can select the appropriate subset of rules (as illustrated in Figure 5.21).

The input molecule also includes a "terminator" sequence that is attached to the right of the input sequence. The input is scanned as long as recognition sites for FokI to cleave still exist and as long as appropriate transition

TABLE 5.7   Analysis of the Effect of Different Spacer Lengths

State Transition	Spacer Length	$\Delta_{current}$	Δ_{new}
$q_1 \rightarrow q_0$	1	0	2
Do not change states	3	0	0
		2	2
$q_0 \rightarrow q_1$	5	2	0

$$GGATGTAC$$
$$GGATGTACCCGA$$
$$T_1{:}q_0 \xrightarrow{a} q_0$$

$$GGATGACGAC$$
$$CCTACTGCTGCCGA$$
$$T_2{:}q_0 \xrightarrow{a} q_1$$

$$GGATGACG$$
$$CCTACTGCGTCG$$
$$T_3{:}q_0 \xrightarrow{b} q_0$$

$$GGATGACGAC$$
$$CCTACTGCTGGTCG$$
$$T_4{:}q_0 \xrightarrow{b} q_1$$

$$GGATGA$$
$$CCTACAGACC$$
$$T_5{:}q_1 \xrightarrow{a} q_0$$

$$GGATGACG$$
$$CCTACTGCGACC$$
$$T_6{:}q_1 \xrightarrow{a} q_1$$

$$GGATGG$$
$$CCTACCGCGT$$
$$T_7{:}q_1 \xrightarrow{b} q_0$$

$$GGATGACG$$
$$CCTACTGCGCGT$$
$$T_8{:}q_1 \xrightarrow{b} q_1$$

FIGURE 5.22 The eight molecular rules for implementing all two-state automata over a two-symbol alphabet. (Courtesy of Ehud Shapiro.)

molecules exist. When the required transition molecules are missing (i.e., no transition molecule exists that matches the current sticky end), the automaton halts and does not reach a final state. If the automaton scans the input string to its end, the final sticky end will be that of the terminator, which reflects the final state reached by the automaton.

To determine whether the automaton accepts the string (i.e., to determine whether the string belongs to the language recognized by the automaton), one has to determine if the final state is an accepting state. The last time FokI cleaves the sequence, it cleaves the terminator sequence, and the resulting sticky end depends on whether the final state is q_0 or q_1.

Output detectors, which are double-stranded DNA sequences with a sticky end that complements the terminator sequences representing either q_0 or q_1, can be used to identify the final state. In Shapiro's method this was done by using output detectors of different sizes for each of the two automaton states. In this way gel electrophoresis, which allows us to ascertain the length of the DNA molecules in the solution, can be used to discover the final state of the automaton.

It is interesting to note that the molecular implementation allows the implementation of both deterministic and nondeterministic finite automata. This nondeterminism is manifested by allowing more that one type of transition molecule to attach to a given sticky end and thus the computation to proceed in alternative routes. Using nondeterminism permits a reduction in the number of states necessary for identifying a particular language relative to a deterministic automaton that recognizes the same language. So it would seem that using the molecular mechanism to implement nondeterministic automata may offer an advantage. However, in practice Shapiro's group noted that increasing the number of nondeterministic decisions decreases the yield (i.e., the number of molecules that complete the computation) exponentially; therefore, this approach was not deemed practical.

5.4 SUMMARY

This chapter demonstrated two main approaches to using biological molecules for computational processes: (1) computations implemented by applying lab techniques to DNA molecules; and (2) independent computation performed by proteins (i.e., the enzymatic computation discussed in Section 5.3). Both approaches depend on choosing an appropriate representation of the data as DNA molecules so that the computation can make use of the complementarity of DNA strands.

DNA computing allows us to harness the inherent power of **parallelism**, as molecular operations occur simultaneously in a huge number of molecules in the test tube. This property is promising for solving computationally hard problems, such as NP-complete problems. Independent computations performed by enzymatic reactions lead to the possibility of using computational processes for medical purposes. For example, systems have been proposed that can identify DNA sequences typical of cancerous processes (i.e., **cancer markers**). Such systems can identify the combination of such sequences that the patient has (e.g., identify the existence of marker A and marker B and the lack of marker C) and can decide on treatment, such as releasing a DNA molecule appropriate for treating a specific condition (Benenson et al., 2004). The treatment might involve turning off genes that promote the cancerous process using various molecular techniques. This approach has been demonstrated in the lab in an *in vitro* setting but is not yet ready for medical use. It is conceivable that such techniques may be used in the future for medical applications.

Note that computations using DNA are "artificial" in the sense that they are not based on natural processes and make use of DNA for applications that are not natural to DNA molecules. Enzymatic computation is based on reactions occurring in nature, but even so the procedure presented by Shapiro in which a sequence of DNA is iteratively digested cannot be regarded as a natural biological process. This raises the question as to whether one can use biological processes and molecules for computing in a manner more similar to their natural activities. This would allow us to make better use of the potential of these molecules, which have been optimized to perform their function over millions of years of evolution.

In **Dennis Bray's** (1995) insightful paper, he claims that a central function of proteins is to transfer information and to perform computations and that proteins are therefore the most useful platform for molecular computations. Particularly useful is the capability of proteins to recognize each other and to attach to each other very specifically. In fact, we can think of the signal transduction mechanism in the cell as a computational process and can harness it for general-purpose computations. **Signal transduction** is the process whereby a combination of signals received by the cell membrane causes a specific chain of reactions to occur (making use of proteins that recognize each other), which results in the expression of genes in the nucleus of the cell. Several such computational models have already been proposed. For example, Unger and Moult (2004) suggested a way (so far only as a theoretical model) to implement logical NAND (not and) gates by molecules built from proteins tagged by DNA sequences. These molecules are diffused in solution and can phosphorylate each other. Phosphorylation is a process by which a phosphate is added to a protein molecule. This phosphate causes the protein molecule to undergo structural modification, in essence creating two versions of the protein—phosphorylated and nonphosphorylated. **Phosphorylation** is a common modification used in biological signaling. Unger and Moult suggested that phosphorylation reactions can implement the logic of NAND gates so that when two molecules collide they create a complex that phosphorylates the target molecule unless they both are already phosphorylated. The model includes additional ingredients required for the model to directly implement any logical circuit, suggesting a way for universal computation by proteins.

In conclusion we can say that the potential of molecular computation of various types is large, yet it is important not to forget that molecular computation has two main obstacles preventing it from currently being a realistic alternative to digital computing:

1. The difficulties in designing molecular algorithms relative to the ease and flexibility of programming digital computers.

2. Physical and experimental limitations making dealing with large, multiphased molecular systems difficult and error prone. Taking a molecular algorithm and turning it into a practical system is far from straightforward.

5.5 FURTHER READING

Adleman, Leonard M. 1994. Molecular computation of solutions to combinatorial problems. *Science* 266, no. 5187, 1021–1024.

Barish, Robert D., Paul W. K. Rothemund, and Erik Winfree. 2005. Two computational primitives for algorithmic self-assembly: Copying and counting. *Nano Lett* 5, no. 12, 2586–2592.

Benenson, Yaakov, Binyamin Gill, Uri Ben-Dor, Rivka Adar, and Ehud Shapiro. 2004. An autonomous molecular computer for logical control of gene expression. *Nature* 429, no. 6990, 423–429.

Benenson, Yaakov, Tamar Paz-Elizur, Rivka Adar, Ehud Keinan, Zvi Livneh, and Ehud Shapiro. 2001. Programmable and autonomous computing machine made of biomolecules. *Nature* 414, no. 6862, 430–434.

Boneh, Dan, Christofer Dunworth, Richard. J. Lipton, and Jiri Sgall. 1996. On the computational power of DNA. *Discrete Applied Mathematics* 71, no. 1, 79–94.

Bray, Dennis. 1995. Protein molecules as computational elements in living cells. *Nature* 376, no. 6538, 307–312.

Lipton, Richard J. 1995. DNA solution of hard computational problems. *Science* 268, no. 5210, 542–545.

Unger, Ron and John Moult. 2006. Towards computing with proteins. *Proteins.* 63, 53–64.

Winfree, Erik, Furong Liu, Lisa A. Wenzler, and Nadrian C. Seeman. 1998. Design and self-assembly of two-dimensional DNA crystals. *Nature* 394, no. 6693, 539–544.

5.6 EXERCISES

5.6.1 Biological Background

1. The DNA polymerase enzyme continues complementing the template strand to its end. Assume we started with long double-stranded DNA molecule and primers that complement regions not at the ends of the molecule. Which molecules will result after the PCR? Hint: draw the molecules, and follow the PCR steps.

2. What will happen if the first phase of PCR is applied to single-stranded DNA molecules? What molecules will result at the end of the process?

5.6.2 Computing with DNA

3. Explain why a graph of N vertices has at most $N!$ paths of length N.

4. Determine whether the graph shown in Figure 5.23 contains a Hamiltonian path (for any v_{in} and v_{out}).

FIGURE 5.23

5. Follow all the steps in Adleman's algorithm, and show that it runs in time that is linear in the number of vertices in the graph.

6. Analyze the probability for errors in Step 4 of Adleman's algorithm where the goal is to select molecules containing a specific subsequence. Let $P(X,s)$ be the "positive" test tube resulting from selecting from a test tube X all sequences containing the subsequence s, and let $N(X,s)$ be the "negative" test tube (i.e., the remaining molecules after taking out all the sequences containing s). Let ε_p be the probability a molecule that should be in $P(X,s)$ ends up instead in $N(X,s)$, and let ε_n be the probability a molecule that should be in $N(X,s)$ ends up in $P(X,s)$ instead. To decrease ε_p, s can be selected by the following repeating selection cycles from the initial population T:

Step 1: $P_1 = P(T,s)$, $N_1 = N(T,s)$
Step 2: $P_2 = P(N_1,s)$, $N_2 = N(N_1,s)$
Step 3: ...
Step 4: ...
Step n: $P_n = P(N_{n-1},s)$, $N_n = N(N_{n-1},s)$
Final step: $P = P_1 \cup P_2 \cup ... \cup P_n$, $N = N_n$

a. Explain the logic of this procedure. Why is there a high probability that the molecules in final test tubes P and N are indeed the correct molecules?

b. What is the probability that after n steps a molecule that should be P is indeed in that test tube?

c. What is the probability that after n steps a molecule that should be in N is in P?

    d. Given an initial estimate of $\varepsilon_p = 1/10$ and $\varepsilon_n = 1/10^6$, what value of n guarantees final probabilities of $\varepsilon_p \approx \varepsilon_n \approx 1/10^6$?

7. Solve the following problem using DNA-based computing: given a map of cities and roads between them where every road has a given length (the map is connected, but not all cities are necessarily directly connected), compute the circular route of shortest total length that visits each city. You can asume that it is possible to distinguish experimentally between circular and non-circular DNA molecules, and sort circular DNA by size.

8. Suggest a way to solve the **vertex cover** problem using DNA-based computation. The vertex cover problem is defined as follows: given a graph, find a minimal subset of the vertices that "covers" all the edges in the graph; that is, every edge has to touch at least one of the vertices in the subset. Refer to Figure 5.24 for an example where the dotted circles indicate the vertices in the minimal cover set.

FIGURE 5.24

9. Solve the **maximal clique** problem using DNA: given a graph, find the largest subset of vertices in which every two vertices are connected to each other by an edge.

10. Show how to use molecular operations to extract only the sequences representing assignments for which C_i is true from the T_{i-1} test tube in the SAT algorithm. Hint: consider as an example the clause $C_i = p \lor q'$. The sequences making this clause true are those where $p = 1$ and those where $q = 0$.

11. Show that the number of operations for the molecular algorithm for SAT is linear in l (the number of literals in the formula).

12. Show that if the number of literals in each clause is constant then the number of operations for the molecular algorithm for SAT is linear in *m* (the number of clauses in the formula).

13. Explain the role of the right frame column in Winfree's binary counter (Figure 5.14).

14. Explain how each of the generic operations in Table 5.3 is carried out molecularly and where each operation is used in the Hamiltonian path and SAT algorithms.

5.6.3 Enzymatic Computation

15. Find the language accepted by the automaton in Figure 5.25.

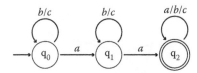

FIGURE 5.25

16. Construct a nondeterministic automaton over the alphabet {*a,b*} that accepts all the words containing *aa* or *bb*. Hint: construct automata for each of the sequences, and combine them using an initial state that guesses which of the two sequences has to appear in the input.

17. Step through the computations performed by the molecules in Figure 5.21, writing down all the partial results.

18. Given the automaton in Figure 5.26, determine the molecules required to represent it, and follow the algorithm's operations on the input *bbaab*.

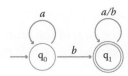

FIGURE 5.26

5.7 ANSWERS TO SELECTED EXERCISES

1. Molecules comprising only the sequence between the two primers will be amplified.

2. There will be no difference from PCR that starts with double-stranded DNA (assuming that both correct primers were used). After the primer attaches to the single-stranded DNA, the DNA polymerase will complete the complementary strand. Then the situation is back to that described in the presentation of the PCR method.

4. The graph in the question does not contain a Hamiltonian path, but note that by adding a single edge we do get a graph (Figure 5.27) containing a Hamiltonian path.

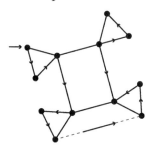

FIGURE 5.27

6. Analysis of the probability using repeated extractions:

   a. Since ε_p is much larger than ε_n the goal is to reduce the probability that a molecule that should be selected is not selected. Thus, at every step we extract the molecules containing s from the remainder of the previous extraction; that is, in every step we select the positive molecules from the negative test tube of the previous step.

   b. The probability is the sum of the probabilities that the molecule is in the positive test tube for each of the n steps; that is,

   $$\left(1-\varepsilon_p\right)\left[1+\varepsilon_p+\cdots+\varepsilon_p^{n-1}\right]=\left(1-\varepsilon_p\right)\frac{1-\varepsilon_p^n}{1-\varepsilon_p}=1-\varepsilon_p^n$$

   (this is a sum of a geometric series).

c. $1 - (1 - \varepsilon_n)^n$.

d. For $n = 6$, $\varepsilon_p = 1/10^6$, $\varepsilon_n \approx 6/10^6$, which is close enough to be considered equal for practical purposes.

7. The solution is similar to the Hamiltonian path solution, with the following changes:

a. As the path is circular, there is no need to distinguish between edges starting at the first vertex or ending at the last vertex.

b. We will add an arbitrary sequence of the length of the edge into the representation of each edge. For example, an edge of length 5 between the vertices i and j is built as shown in Figure 5.28.

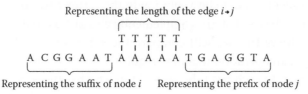

Representing the length of the edge $i \rightarrow j$

T T T T
| | | |
A C G G A A T A A A A A T G A G G T A

Representing the suffix of node i Representing the prefix of node j

FIGURE 5.28

c. Mix as before all the edges and the complements to the vertices to create all the circular paths. Note that, since the first and last edges are no longer distinct, all possible circular paths will be generated. Note that circular DNA runs in a particular way on a gel (circular DNA migrate more slowly on electrophoresis gels and their migration rate is determined by their radius of gyration), a fact that can be used to extract only circular molecules.

d. As in Adleman's algorithm, we will use n test tubes to extract the sequences containing the complementary sequences to the sequences representing all the cities.

e. The remaining molecules (i.e., those containing all the vertices) will undergo gel electrophoresis, and the shortest molecule will be selected as it represents the circular path of minimal length.

8. Represent the given graph's vertices and edges as in Adleman's algorithm. Then create sequences representing all subsets of the set of vertices. This can be achieved using a graph similar to the one we used when solving SAT where each vertex is represented in the upper edges and one dummy vertex in all the lower edges (Figure 5.29). The total length of all dummy nodes should be less than the length of the representation of a real vertex. This will produce sequences containing all the subsets of the vertices, where all missing vertices from the set are represented by the dummy vertex. We are interested in only the subsets whose vertices cover the whole graph, that is, in which every edge touches at least one vertex in the subset. Recall that the representation of each edge contains half of the representation of the two nodes it connects. Thus, for a sequence representing a subset of nodes to be a solution to the vertex cover problem it must contain bases that are complementary to at least half of each edge. To achieve this we will compare the subsets with all edges in the graph, an operation that may require n^2 test tubes for a graph with n vertices. In each test tube we will test whether the edge hybridizes to the sequence (we can create the representations so that the hybridization occurs if half of the edge's length matches the sequence). At the end of the process the remaining sequences represent covering subsets. By performing gel electrophoresis we will identify the shortest sequences corresponding to minimal cover set.

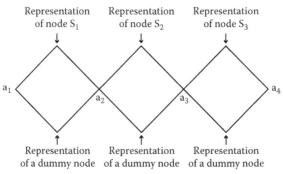

FIGURE 5.29

10. Let $C_i = p \vee q'$. Represent C_i as $v_1 \vee v_2$, where v_1 is the variable p, and v_2 is the negation of the variable q.

a. Since v_1 represents a variable, let t_1 be a test tube containing all the sequences extracted from T_{i-1} where v_1 is assigned the value 1. Extraction can be done using magnetic beads that are complementary to sequence of v_1. Let $\bar{t}_1$ be the reminder of sequences left in T_{i-1}.

b. t_2 is created from sequences in the reminder test-tube, $\bar{t}_1$. Since v_2 represents a negation of a variable, select from $\bar{t}_1$ all the sequences where v_2 is assigned the value 0, that is, that include the q' sequence.

Now mix together the contents of the test tubes t_1 *and* t_2. This creates the T_i test tube because in the first step we extract all sequences satisfying v_1. From the remainder we extract all sequences satisfying v_2, so the result is all the sequences satisfying v_2 but not satisfying v_1 (it is trivial to generalize this example to clauses containing any number of literals). The mixing step gives us the union of the sets, so we end up with all the sequences satisfying v_1 or v_2. (If the clause contains three literals like in 3SAT problem we need to create in a similar way a third test tube t_3 and mix the three test tubes to create T_i.)

11. The first step of the algorithm requires the preparation of raw material (representations) that depends linearly on n (calculate the exact number of required molecule types), followed by mixing the molecules and waiting for the ligation to finish. The next steps require a number of extraction operations that is linear in the number of literals (Step 2) and one identification step to test whether any DNA remains in the test tube (Step 3).

13. The role of the right frame column is to add 1 to the value computed in the previous row; therefore, it is implemented using a tile with a protrusion causing the computation in the next row to start with a carry.

16. See Figure 5.30.

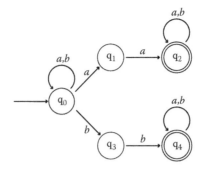

FIGURE 5.30

18. The four rules needed to implement the automaton are as follows:

$$q_0,a \rightarrow q_0$$

$$q_0,b \rightarrow q_1$$

$$q_1,a \rightarrow q_1$$

$$q_1,b \rightarrow q_1$$

Thus, we need to use molecules T_1, T_4, T_6, and T_8 from Figure 5.22.

CHAPTER 6

The Never-Ending Story

Additional Topics at the Interface between Biology and Computation

I N THIS BOOK WE have discussed a wide spectrum of ideas in computer science that were inspired by our understanding of biological processes. We focused on four main areas that introduced new **computational models** based on ideas and insights arising from biological research. Chapter 2 dealt with **cellular automata** in which computation is performed on a grid of cells, and every cell affects only its neighbors. This model is somewhat reminiscent of a colony of single-cell organisms (e.g., bacteria), which presents a complex collective behavior, even though each cell's behavior is based on a set of relatively simple local rules. We saw how to prove that a nonstandard computational model is universal (in the sense of being equivalent to Turing machines). We also saw how cellular automata allow us to formally study the conditions that are sufficient for self-replication, a fundamental aspect of living systems.

Chapter 3 dealt with **evolutionary computation**, which involves solving computational problems such as optimization and search problems by mimicking the evolutionary process in nature. The focus of Chapter 3 was mainly on **genetic algorithms**; even this specialized model presents the system designer with a wide scope of choices, such as how to represent the genetic data or the precise properties of the genetic operators. Moreover, we saw that one can formally prove theorems about properties of the evolutionary computation (Holland's Schema theorem) that are valid under a wide variety of assumptions. We also discussed **genetic programming** in which the individuals undergoing evolution are representations of computer programs.

Chapter 4 presented several models of **neural networks**. These models are based on an attempt to mimic the way a brain operates in order to facilitate **machine learning**. The first three models were based on **supervised learning**, where networks are presented with a set of examples for which the expected output is known, which are used to train the network. The first model was that of a **simple perceptron**, and we saw how simple it is to prove the inherent limitations of such networks. Then we discussed **multilayered neural networks** and developed a learning algorithm based on **backpropagation**, which adjusts the network weights automatically and sequentially. Then, using the **Hopfield model**, we studied the issue of associative memory and discussed the strength and the weaknesses of this model. Finally, we gave an example of **self-organizing maps**, which are neural networks capable of **unsupervised learning**.

Chapter 5 dealt with **molecular computation**. Here we saw another kind of link between computer science and biology: the use of biological techniques and organic molecules to implement computational processes. In this chapter the biological material was used as "hardware." The inherent parallelism of molecular processes was harnessed to improve the efficiency of expensive computations such as finding Hamiltonian paths in graphs. We showed how DNA can be used as a computational medium as well as how to harness enzymatic reactions to implement autonomous computational processes.

Each of these four areas is an extensive field of study, and we presented only basic concepts—enough to give a sense of each topic to help the reader identify which approaches may be useful for solving a given problem, and to provide the basic tools and concepts needed for the further study of each topic. The reader interested in more in-depth knowledge will find references to further reading at the end of each chapter and in the list of recommended books that appears at the end of this chapter.

The summary of the previous chapters shows how combining ideas from biology and computer science leads to a variety of results: from technological and engineering applications to theoretical conclusions and formal proofs of theorems. In this chapter we will discuss briefly additional topics that are in the intersection of computer science and biology that further demonstrate the rewards that can come from "cross-fertilization" between the two fields. We will try to relate the new topics to ideas discussed in the previous chapters and to highlight the similarities and differences.

We will not discuss the important field of **bioinformatics**, which focuses mainly on computational analysis of biological sequences such as RNA, DNA, and proteins. This analysis is achieved to a large extent by using methods developed in the computer science fields of pattern matching and machine learning. This is a central topic in computational biology requiring a separate discussion, and many recent textbooks have been dedicated to this subject (see Further Reading section).

6.1 SWARM INTELLIGENCE

Swarm intelligence is a set of computational approaches influenced by observing the living world from a computational perspective that views the behavior of organisms as problem-solving processes. Swarm intelligence approaches are derived from observations showing how cooperating organisms solve problems collectively. The classic example of such behavior is that of an ant colony. Anybody who has observed a row of ants marching toward a food source must have wondered how the ants know where to go and how to return to their nest. It turns out that an ant that has discovered a food source can signal the preferred direction to other ants, which repeat the process and mark the way for even more ants.

Swarm intelligence is based on the observation that colonies of simple organisms can present a behavior that seems planned and goal oriented even though each individual is simple and lacks the skills to solve the problem independently. We have discussed this observation in other contexts, particularly when discussing the emergent behavior of cellular automata in Chapter 2. In this respect, the discussion can be considered as an extension of the topics described there. The set of simple organisms that are capable of developing mutual interactions and interact with the environment is known as the **swarm** or the **swarm system**. The collective goal-directed behavior is an **emergent property** (a topic we return to in Section 6.3) of the swarm system and is referred to as **swarm intelligence**. We present three computational methods based on swarm intelligence.

Ants leave chemical markers called **pheromones** on their trails, which allow them to pass information between individuals. Communication between individuals by locally changing the environment is called **stigmergy**. The scent of the pheromones is picked up by the olfactory organs of other ants, allowing the pheromones to mark the way back to the nest and the way to food sources. When more ants use a particular path, the stronger its markings will become. Conversely, pheromones evaporate over time, so a trail that has been neglected will disappear after a certain period of time. Note that ants do not always follow the trail and may turn in random directions from time to time. This guarantees that ants will discover new food sources and new trails.

The combination of these two properties—individuals with simple computational abilities and communication using the environment—makes swarm computation appropriate for distributed computing and for solving coordination problems for robots that operate cooperatively.

6.1.1 Ant Colony Optimization Algorithms

The pheromones mechanism gave rise to the development of a class of optimization algorithms known as **ant colony optimization (ACO) algorithms**. Such an algorithm was first presented by **Marco Dorigo** in the 1990s (Bonabeau et al., 1999, 2000). The goal of the algorithms is to find an optimal solution to a computational problem by using the method that allows ants to find food quickly and efficiently by not wasting energy on long trails. The problems are usually presented as finding good paths in graphs, and the algorithm proceeds by creating a set of virtual ants that walks the graph with the goal of constructing appropriate paths. A typical application is the **traveling salesperson problem** (TSP). The solution to this problem is the shortest path traversing a given set of cities where each city is visited exactly once.

To solve the TSP, several ants are placed in each city. At each time step, a random ant is selected and has to travel on the graph according to the trails marked by pheromones. Thus, the probability that an ant will go to an adjacent city (an adjacent node in the graph) is directly proportional to the amount of pheromones deposited on the edge between the current and adjacent cities.

The following **random proportional transition rule** is the formula one of the algorithms uses to determine the probability that an ant at vertex i will go to the adjacent vertex j:

$$p_{i,j} = \frac{\left(\tau_{ij}^{\alpha}\right)\left(\eta_{ij}^{\beta}\right)}{\sum_{k}\left(\tau_{ik}^{\alpha}\right)\left(\eta_{ik}^{\beta}\right)}$$

where τ_{ij} is the amount of pheromones on the edge (i,j), and η_{ij} is the heuristic value assigned as the value of the edge (i,j) a priori. This heuristic value serves to estimate the quality of the ants' choices in advance of building

the path. Such heuristic functions exist also in algorithms for game playing where the different game states are evaluated as part of evaluating the
game tree. In problems such as TSP the heuristic value is determined by
the distance between the cities d and may be, for example, $1/d$, to give a
higher weight to closer cities. As expected, the heuristic value is computed
using local information only. α is a positive constant that determines how
the quantity of pheromones influences the algorithm, and β is a positive
constant determining the influence of the heuristic value. The sum in the
denominator is over all the neighboring vertices k among which the ant
has to choose.

After building a path between n nodes, the ant updates the amount of
pheromones on the graph edges it traversed in accordance with the quality of the complete path (the better the path, the more pheromones will
be deposited). For simple graphs, this mechanism suffices for finding the
shortest paths. For complex graphs or when searching for paths with other
properties, additional mechanisms are added to the algorithm, such as
having only the ant that found the best path during an iteration of the
process leave a pheromone trail behind it; having a certain percentage of
pheromones evaporate at every time step; fixing minimum and maximum
values for the amount of pheromones deposited on each edge; or keep-
ing a list of visited cities for each ant to avoid multiple visits to the same
city. Researchers have successfully used ACO to solve the TSP and other
similar combinatorial problems, which can be represented as problems of
finding paths in graphs.

Another interesting use of swarm intelligence is for planning routing
tables in a communication network. The basic premise is simple: communication packets update the routing tables based on the quality of the path
they were routed to. For slow routes, the corresponding table entries will
be updated by a small value, whereas a faster path will be updated with a
higher value. In this fashion the packets act as ants leaving a trail of pheromones behind them as well as carrying the information in the network.
This application of swarm intelligence is called **ant colony routing** (**ACR**)
and has two important engineering advantages:

1. Using the many packets sent in the network allows for an efficient
   mapping of the network which may be large and complex.

2. The mapping happens in real time and allows for route changes
   based on the changing characteristics of the network. Since the load

in certain parts of the network may vary greatly over time, this property is highly important.

6.1.2 Cemetery Organization, Larval Sorting, and Clustering

Observing ant behavior also leads to techniques for solving **clustering** problems. In these problems the goal is to find a good partition of a (usually large) set of data into subsets or clusters. The goal is to have the elements of each cluster be closer or more similar to each other than to members of other clusters. The number of clusters may be an input to the algorithm or may be determined by it. A typical example is the problem of partitioning customers into sets of customers with similar characteristics, for example, "customers who buy expensive kitchenware" versus "customers who buy expensive appliances and cheap kitchenware." The types are not known a priori—they are found by the algorithm that attempts to identify clusters minimizing the distance between the data points inside each cluster.

Observation of certain species of ants has demonstrated that they arrange "cemeteries" for dead ants in the nest. Initially, the dead ants are distributed randomly over a certain area, but after some time the area is partitioned into subregions containing dead ants and others that are free of them; in other words, one can observe clusters of ant carcasses. Other ants have been shown to arrange their larvae by size, such that the smaller larvae are placed in the center of a cluster and the larger larvae at the cluster's periphery. These observations gave rise to the idea of using stigmergy to solve clustering problems similarly to the ants' techniques for clustering carcasses. Here the data to be partitioned and sorted play the role of the "carcasses" to be clustered.

The basic idea of the algorithm is to place the data on a two-dimensional grid, similar to the one found in cellular automata. For best results the data are initially placed on the grid randomly using a uniform distribution. Ants are also placed on the grid and may move from one grid cell to another and carry with them the data when they move. At each time step an ant decides whether to move the datum in the cell according to the distribution of data in the local neighborhood, and it may do so only if it is not already carrying some other data. The sparser the data in the neighborhood, the higher the probability the ant will "pick up" the datum and start carrying it. Conversely, an ant may put a data item it carries down in a new cell at any time step, and the probability of this event increases the more data items there already are in the local neighborhood. Iterating this process over a number of generations causes the data to cluster, as more distant data are brought into existing clusters.

When calculating the probability of an ant picking up or putting down data, one has to take into account both the placement of the data on the grid and the distance between the datum in the current cell and the data in the local neighborhood. Going back to our example, the distance between two customers might be defined as the number of items one customer has bought and the other has not (i.e., the size of the symmetric difference between the sets of items each customer has bought). Using this definition, customers who bought similar items will be considered "close" for the purpose of clustering. Note that in most cases the data are characterized by a rather large set of properties, so they can be considered as points in an n-dimensional space R^n (where n is the number of properties for each data point). The algorithm we described not only attempts to discover clusters but also does so while projecting the n-dimensional property space onto the two-dimensional grid (or more generally on a space with less than n dimensions). It is convenient to implement this approach using "ants" moving on a discrete grid, so in fact the n-dimensional space R^n is projected onto a discrete two-dimensional space (Z^2) similarly to what is achieved by self-organizing maps (Chapter 4).

The structure of the clustering algorithm is as follows:

```
// Generic code for clustering using ants (Lumer-Faieta Algorithm)
// Each ant remembers its current location on the grid, and the item
// it is carrying.

PLACE_ITEMS_ON_GRID()

PLACE_ANTS_ON_GRID()

WHILE not END_CONDITION
  BEGIN
        FOR i:=1 TO number of ants
        BEGIN
          IF not(CARRYING(ant_i)) and not(EMPTY(LOCATION(ant_i)))THEN
              p := PICKUP_PROBABILITY(ant_i)    // see below
              PICKUP(ant_i) with probability p
          ELSE IF CARRYING(ant_i) and EMPTY(LOCATION(ant_i)) THEN
              p := PUTDOWN_PROBABILITY(ant_i) // see below
              PUTDOWN(ant_i) with probability p
          END IF
          MOVE(ant_i)   // randomly move ant
        END
  END
```

To complete the presentation of the algorithm we have to specify how to compute the probability that an ant will lift up a piece of data and the probability that an ant will deposit a data item at a particular grid point. These probabilities have to be based on the number of similar points in the ant's neighborhood.

First, we define the function $f(i,r)$, which computes the *local density* of objects similar to object i located at position r:

$$f(i,r) = \begin{cases} \dfrac{1}{s^2} \sum_j \left[1 - \dfrac{d(i,j)}{\alpha} \right] & \text{if positive} \\ 0 & \text{otherwise} \end{cases}$$

where $d(i,j)$ is the distance (or dissimilarity) between objects i and j, s is the neighborhood's radius, and the sum is over all items in the neighborhood around r. $f(i,r)$ measures the average similarity between item i and the items in its neighborhood. The parameter α determines the sensitivity of the comparison: if the value of α is high, then the comparison is less sensitive and items that differ by much may be clustered together. Conversely, when α is low, even similar items will be viewed as different and will not be clustered together.

Using $f(i,r)$ we can define the probabilities for picking up and putting down data using the following formulas:

$$Pickup\_Probability(ant_i) = \left(\frac{k_1}{k_1 + f(Item(ant_i), Location(ant_i))} \right)^2$$

$$Putdown\_Probability(ant_i) = \begin{cases} 2f(Item(ant_i), Location(ant_i)) & \text{if } f(\text{Item,Location}) < k_2 \\ 1 & \text{if } f(\text{Item,Location}) \geq k_2 \end{cases}$$

where k_1 and k_2 are constants. If $f(i,r) \ll k_1$, the pickup probability will be close to 1. This describes a sparse neighborhood. If $f(i,r) \gg k_1$ the neighborhood is rich in similar items, and the pickup probability will be close to 0. k_2 plays a similar role in computing the putdown probability.

To summarize, the probability of a pickup decreases with the density of similar items, whereas the probability of an item being put down increases with the density of similar items in the neighborhood. In this way the algorithm achieves the desired goal.

We saw in previous chapters how the interaction with the environment can often affect the behavior of the computational organism. It is interesting to note that this is also the case with swarm intelligence: the system is governed by the feedback the ants receive from the environment, which is used both for representing the input to the algorithm and for communication between the organisms in the swarm.

6.1.3 Particle Swarm Optimization

Swarm behavior—particularly the behavior of schools of fish and flocks of birds—has led to another computational technique, called **particle swarm optimization (PSO)**. In this technique optimization problems are solved by a set of particles distributed on the search space (which is represented as an n-dimensional space R^n) where each point (an n-tuple of real numbers) represents the n characteristics of a possible solution. The particles move around attempting to reach the extreme points (optimal solutions) identified at each time step. In the basic algorithm, every particle is aware of the following:

- The quality of the solution represented by the point at which it is located.

- The quality and location of the best solution it has ever visited (*personal best*).

- The quality and location of the best solution the population has ever encountered (*global best*).

The particles are initially randomly distributed on the search space, and each has a velocity (which initially may be 0 or some random value). At each iteration the location and velocity of the particles are updated as follows:

1. The change in velocity ("acceleration") is determined so that it creates movement toward the personal best and global best. The acceleration is computed as a weighted average of the distance between the particle and the personal best and global best; the location and velocity of particles are vectors in the n-dimensional space (i.e., vectors

of n real numbers). The new velocity is calculated using the current velocity and acceleration.

2. The new position of particle i is determined by the current position and the new velocity ($\mathbf{x}_{i,t}$ is the vector location of particle i at time t): $\mathbf{x}_{i,t} = \mathbf{x}_{i,t-1} + \mathbf{v}_{i,t}$. (Figure 6.1).

As usual, the algorithm executes until it converges or some other halting condition is satisfied.

Most of the parameters allowing for fine-tuning the system involve the way the new velocities of the particles are calculated in Step 1. The new velocity of particle i is determined as follows:

$$\mathbf{v}_{i,t} = \omega\mathbf{v}_{i,t-1} + c_1 rand()\left(globalbest - \mathbf{x}_{i,t-1}\right) + c_2 rand()\left(personalbest_i - \mathbf{x}_{i,t-1}\right)$$

where:

- $x_{i,t}$ is the location of the particle at time t.

- *globalbest* is the location of the best solution the population has encountered so far.

- *personalbest* is the location of the best solution particle i has encountered so far.

- ω denotes the *inertia* of each particle. It is usually chosen to be close to 1.

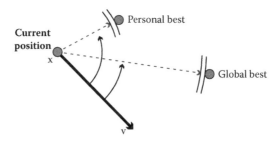

FIGURE 6.1　New position calculation. The actual trajectory of a particle is determined by its own position and velocity and also by biasing the trajectory toward the best position visited by the particle and the best position found by the entire population.

- The constants c_1 and c_2 determine the influence of *personalbest* and *globalbest*, respectively. The larger c_2, the larger the influence of the particle's "private" data, whereas the larger c_1, the larger the influence of the whole population on the behavior of each particle (the *social influence*). Initially, one could use $c_1 = c_2 = 2$.

- *rand()* is a random number between 0 and 1. Note that a different random value is used for each dimension (to simplify notation this is not reflected in the formula above).

The calculation of the velocity in each dimension is done separately and takes into account the corresponding dimensions of *globalbest* and *personalbest*.

Note that, in contrast to the cellular automata described in Chapter 2 where each cell has only local knowledge, in PSO, the system keeps track of the global data—the location of all particles and the properties for the optimal solution *globalbest*. It is possible to reduce the role of global information by having each particle be aware only of a limited group of neighboring particles, where the neighbor relation is defined in advance and does not depend on the current locations of particles.

The structure of the PSO algorithm is as follows:

```
// Generic code for implementing a simple PSO algorithm
// The algorithm tries to maximize the fitness function f()

INIT_POPULATION()   // Create initial population
WHILE not END_CONDITION
 BEGIN
      FOR i:=1 TO number of particles
        BEGIN

        // Remember location of personal best

        IF f(x_i)>personal_best_i THEN
           personal_best_i := x_i
        END IF

        // Remember global best

        global_best:=
             MAX_FITNESS_LOCATION(personal_best_1,personal_best_2,...,
                                  personal_best_n)

        FOR d:=1 TO number of dimensions
          BEGIN
          // notice that in practice we only need to store all the
          // current positions, and all the new positions (two sets of
          // locations), not all the previous values.

             v_{i,t} := ωv_{i,t-1} + c_1 · rand()· (global_best - x_{i,t-1}) +
                                     c_2 · rand()· (personal_best_i - x_{i,t-1})

             x_{i,t} := x_{i,t-1} + v_{i,t}


          END  // FOR d
      END  // FOR i
 END
```

In an impressive demonstration, particle swarm optimization was used to find weights for a neural network whose goal was to report the charge of an electrical car's batteries. The network consisted of five input neurons, three hidden neurons and one output neuron, and it took around 3.5 hours to train using backpropagation. It took merely 2.2 minutes to find weights achieving the same level of success (the same sum-squared error) using PSO (Kennedy and Eberhart, 2001, p. 318).

6.2 ARTIFICIAL IMMUNE SYSTEMS

Immunology is the research field dealing with the immune system, which defends the organism against a wide variety of pathogens such as bacteria, fungi, and parasites. The immune system is a complex system whose description is beyond the scope of this book. We will present a few applications that employ insights gained from knowledge of the biological immune system to solve problems arising in computer science and while doing so will introduce the relevant properties of the biological immune system.

The immune system is fascinating from a computational perspective as it operates in a consistent fashion, "reaches" conclusions, exhibits "memory," and performs various activities while being totally distributed and without a central control mechanism or even "wiring." Thus, it is radically different from the nervous system discussed in Chapter 4. Two central properties of the immune system are **immune specificity** and **immune memory**. Immune specificity refers to the capability of certain immune system cells to identify specific pathogens, to target them, and to destroy them. Immune memory is based on the fact that some of the cells generated during the initial contact with a pathogen remain in the organism and allow for a faster reaction to subsequent attacks by the same pathogen (this property is the basis of vaccinations). Artificial immune systems, which mimic biological immune systems, attempt to recreate these properties to achieve computational needs. We will describe one such task—securing a computer system against unauthorized users.

To achieve its immunological task, the immune system has to distinguish between elements belonging to the organism and external elements. This is called the distinction between *self* and *nonself*. To achieve this goal the system has to contain cells that recognize and react to new elements invading the system. This raises an interesting question: how can one create cells that can identify elements the organism has never encountered previously? We can conceive of various ways for creating detectors for certain known properties, but how can one detect invaders with unknown properties? Another requirement from the immune system, which is of equal importance, is that the detectors do not react to the organism's own cells (the self), because such a reaction will cause the organism to attack itself (**autoimmune** diseases, such as multiple sclerosis and lupus, occur due to an immune response to the organism's own proteins). It turns out that even healthy people have autoimmune activity but to a lesser extent. This raises the question of whether the immune system is geared only toward identifying the nonself or whether it has other functions unrelated to dealing with external elements. For instance, it might have additional regulatory functions. This fundamental question gives a completely different perspective to the immune system's function, whereby the immune system actually *defines* and *maintains* the *self* rather than just *identifying* it. In our artificial applications we will ignore this question and focus on the task of distinguishing the self from the nonself.

One way the immune system attempts to distinguish between self and nonself is **negative selection.** A large set of detectors are created randomly, and the ones that react to the *self* are sieved out. To implement this, many detectors are generated and allowed to live for a certain period of time in an environment exposed to the organism's own molecules. If the detectors fire, there is a high probability they are reacting to the organism itself and have to be removed. Detectors that have not fired during this training period have the potential to react only to external elements and therefore should be activated in the hope that they identify attackers. In this way the system can be said to "learn" to identify nonself elements. In the immune system this process happens mainly in the thymus gland where white blood cells known as **T-cells** are "trained" to distinguish between self and nonself targets. It is easy to see that this method is useful for identifying not only attackers but also any anomalies in a system.

This process is called negative selection because the detectors reacting against the organism's own proteins are removed. This is not the only mechanism employed by the immune system, and other mechanisms have inspired various learning algorithms; however, in this section we will focus only on negative selection.

6.2.1 Identifying Intrusions in a Computer Network

It seems very natural to use ideas derived from immunology to defend computer systems against unwelcome intruders, as the biological immune system attempts to solve a similar problem. The **LISYS** system we will describe (Hofmeyr and Forrest, 2000) is based on the negative selection mechanism. In addition, the system makes use of other ideas inspired by the immune system.

The goal of the system is to defend a local area network (LAN). The system monitors the communication in the network constantly and learns to distinguish between normal communication (self) and unusual communication (nonself). The system makes use of the fact that in a LAN every computer sees the entire communication passing in the network. This allows for detectors to be distributed on many computers on the network and for all the communication passing through the network to be monitored from each one of them.

The monitoring system observes the network connections between different computers. Each connection is represented by a 3-tuple composed of the Internet Protocol (IP) address of the sending computer, the IP address of the receiving computer, and the requested service (the port). Each 3-tuple is represented by a string of 49 bits. The goal of the system is to distinguish between 3-tuples representing normal connections between computers and those that are atypical and may indicate unauthorized entry into the network. For this purpose the system has to compare strings representing detectors with strings representing active connections between two computers. The strings are compared using the *r-contiguous bits* criterion, which considers two strings as matching if there exists in both strings an identical contiguous substring of at least *r* bits.

Negative selection plays a role in the creation of new detectors. A detector, which like the network connections is represented as a string of 49 bits, is generated randomly. The detector is considered *immature* during a training period of length *T* called the **tolerization** phase. If during that time the detector fires, the assumption is that it reacts to *self* strings and the detector is eliminated. A detector surviving this initial phase is considered *mature* and is used to identify invaders. A mature detector identifying at least τ strings in a time interval is considered to have identified an invasion, and its state changes to *active* (and its *match counter m* is reset to 0). τ is called the *activation threshold*. The match counter decays with time, so if not enough strings are identified during a time interval the detector slowly reverts to a less active state.

When a new 3-tuple is observed it can cause a few detectors to fire. Those identifying an intrusion best (i.e., with the largest number of adjacent identical bits with the 3-tuple) are selected to be *memory detectors*. These detectors clone themselves, and the clones are distributed to neighboring computers in the network. In this way the identity of the atypical 3-tuple is distributed in the network and it will be quickly identified at its next occurrence as detectors identifying it will exist on many computers in the network. Moreover, the activation threshold of memory detectors is lower than that of regular detectors (e.g., $\tau = 1$), which causes them to react faster. The memory detectors provide the intrusion detection system with an **immune memory**.

Similar to a biological immune system, one of the greatest dangers is of the system being overly sensitive and reacting strongly to innocent occurrences. A way of minimizing this is by **co-stimulation**, which involves generating an immune reaction only when a number of different mechanisms detect a problem. The LISYS architects chose a simple method for co-stimulation: when a detector is activated by identifying a string s, the string is sent to a human operator, who has to confirm to the detector within a fixed time T_s that this indeed is an abnormal occurrence. Only then does the detector become active and an immune response is initiated. If the operator does not respond within this time period, the assumption is that the detector identified a valid string (self), and the detector is removed from the system and replaced with a new immature detector. The life cycle of a detector is shown in Figure 6.2. The system developers tested the immune system on a set of data obtained from a live communication network by stimulating 20 days of real network use. Tables 6.1 and 6.2 list the immune system parameters they chose and the performance achieved by the system.

6.3 ARTIFICIAL LIFE

The borderline between using biological ideas to solve computational problems (**bio-inspired computing**) and attempting to build systems that behave like biological organisms (**artificial life**) is fine and often hard to define. Throughout the book we have mainly addressed bio-inspired computing, but we will now attempt to differentiate between various approaches to artificial life (**ALife**) and will discuss a few well-known systems.

Discussing artificial life immediately raises the question of defining what life is in an exact way (the **definition of life** problem). Life manifests itself in a vast number of different living organisms with their different properties. Our large but limited knowledge of biological systems and the philosophical depth of the question "what is life" combine to make the definition of life a question many scholars prefer to avoid. The scientists and philosophers who did discuss the problem have suggested an array of definitions focusing on the many properties found in living organisms.

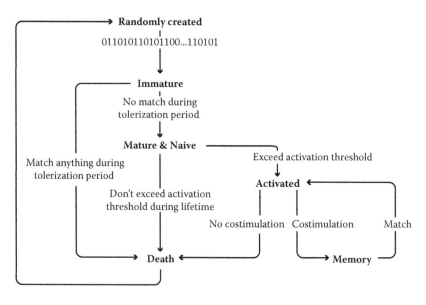

FIGURE 6.2 The life cycle of a detector. (Adapted from Hofmeyr, Steven A. and Stephanie A. Forrest, *Evolutionary Computation* 8, no. 4, 443–473, 2000. With permission.)

TABLE 6.1   LISYS Parameters

Parameter	Value
String length	49 bits
Number of contiguous bits to match (r)	12 bits
Activation threshold (τ)	10 matches
Decay period of match counter (m)	1 day
Decay period of local sensitivity (see Exercise 6.13)	0.1 days
Telorization period of immature detector (T)	4 days
Waiting period for costimulation (T_s)	1 day
Detector life expectancy	14 days
Number of detectors per node	100 detectors

TABLE 6.2   LISYS Performance

Percentage of immature detectors in detector population (average over 20 days)	23%
Average number of false positives per day	1.76
Number of correctly identified intrusions into system	7 of 7

Among these properties we may find self-replication and heredity, adaptability to the environment and homeostatis, and metabolic behavior (i.e., the capability to use matter and energy in the environment for the organism's existence and functioning).

The challenge of providing such definitions is twofold: (1) the definition has to match all the varied objects we consider as living organisms and not match objects such as chairs, rocks, and digital computers that we do not consider to be live organisms; and (2) we have to avoid using too narrow a definition that will match only the living organisms found on Earth. We should aim to reflect on the fundamental properties of life, so that whenever we encounter an object that falls within the definition, we will agree it is alive, whether it is on another planet or even on a computer system. Due to these and other difficulties, there is no one definition everyone agrees on, and it is doubtful that such a definition is possible. Researchers who work in the field of artificial life do not necessarily define the term explicitly, but we can nonetheless characterize the assumptions underlying many of their projects.

The fundamental premise underlying artificial life is that life is not limited to phenomena we necessarily know already. Artificial life researchers thus deal with questions of **life-as-it-could-be** as well as **life-as-we-know-it**. This means that man-made systems are not ruled out (at least in principle) as living systems. The reasoning behind this assumption is that life is a dynamic process with universal characteristics that are independent of the life's medium. In other words, life is a characteristic of the way the medium is organized and not of the medium itself: for instance, life does not have to be based on organic molecules. This allows us to accept the possibility that a computerized simulation of living processes should be considered as being alive.

Given the scientific and philosophical difficulties of defining life, it is common among researchers to distinguish between two types of approaches to artificial life: (1) the **strong ALife** approach, which postulates that virtual "creatures" on a computer screen can be considered to be alive if they fulfill the definition of life used by the researchers; and (2) the **weak ALife** approach, whereby computerized creatures displaying characteristics of living systems are only *models* used in research and are not really alive.

Most ALife systems have common characteristics, based on general ideas derived from biology and the study of complex systems. As expected, these characteristics are similar to the main characteristics of the biology-inspired models which we have discussed throughout this book:

1. The systems are composed of a large collection of simple programs or other simple entities (a *"population"*).

2. There is no central control mechanism.

3. Every program or object reacts to local phenomena in its immediate environment. The environment may of course contain other objects with which the object has to interact.

4. Any property of the system that extends beyond the local behavior of the objects (i.e., an **emergent property**) is the result of the local simple behaviors.

As expected from this list, many ALife systems use the computational models described in Chapters 2, 3, and 4.

We will describe a few representative examples of ALife systems. While studying them, try to determine how well these examples adhere to any of the previously given characteristics and whether they deal with the challenge of strong ALife or the simpler but still challenging weak ALife.

6.3.1 Avida

Avida (Lenski et al., 2003) is a software environment for studying and evaluating the evolution of self-replicating computer programs. (Following the lead of the system's developers, we will call such programs **computational organisms.**) Using Avida allows researchers to perform experiments on artificial evolutionary processes relatively easily and to follow all the stages in the evolution of the computational organisms "living" in the computer's memory. Computerized experiments are of course simpler than experiments in the laboratory, especially when the experiments involve following many generations of organisms. Many researchers use mathematical simulations to analyze evolutionary processes, but this approach is inherently biased as the simulations are based on the researchers' already held assumptions about the evolutionary process (the computer simulations simply use pseudo-random numbers to explore probabilistic models of evolutionary processes). The Avida developers chose a different route—the computer is used not to perform the computations defined by a mathematical model but as an environment in which autonomous organisms operate. These organisms are responsible for their own reproduction and interaction with the environment and create an evolutionary process that does not necessarily operate according to a predefined mathematical model. The main task of an Avida organism (which is a computer program) is to generate as many copies of itself as possible (i.e., to self-replicate). Note the fundamental difference between this approach and the way the genetic algorithms we studied in Chapter 3 work. In Avida the organism is responsible for its self-replication; replication is not provided by a separate mechanism. The success in self-replication is the fundamental metric for an individual's fitness in Avida, and the success of an organism is measured by the number of its copies in the final population.

The Avida software system implements a virtual computer and operating system, on which different programs comprising the population of organisms in the virtual environment are run. The computer runs as a parallel system, using *time slicing*, so that every artificial organism is allocated a time slice during which the computer program comprising the organism is executed. Avida was inspired by an earlier system for studying artificial evolution of self-replicating entities called **Tierra**, which was developed by the ecologist **Thomas S. Ray**. One of the main differences between Avida and Tierra is that in Avida organisms may be assigned computational tasks, and if they fulfill them successfully they are rewarded with extra running time as a bonus. For example, we may challenge the organism with the task of accepting two numbers as input and of producing their sum as output upon completion of the program. Organisms that are successful at this task will get extra running time, which they can use to create copies of themselves. In this fashion one can study the evolution of different computational capabilities. For example, one of the experiments conducted using Avida was to compute logical operations, and the highest bonus was given to the organisms implementing the EQU operation, the operation that tests whether the bits of both input strings are equal (see Lenski et al., 2003). The goal of this work was to study how the evolution of complex traits depends on the evolution of simpler building blocks. Another feature Avida added to Tierra is that it implements a two-dimesional universe on which the organisms live. It additionally supplies a large spectrum of configuration and monitoring mechanisms for evolution of the computational organisms.

Ray already observed interesting evolutionary phenomena using Tierra (Ray, 1992). In his first experiment the only success criterion was the rate of self-replication. Ray noticed that the organisms became successively shorter—since the shorter the program the less time it needs to replicate—therefore for a fixed execution time, shorter programs will generate more copies. Another interesting phenomenon Ray discovered was that some organisms succeeded so well in decreasing their size that they removed critical parts of themselves and used parts of other programs that resided in the computer's shared memory. This is reminiscent of the biological phenomenon of **parasitism**, whereby a parasite benefits at the expense of another organism. The creation of parasites led to an **arms race**: the abused organisms developed methods to confuse the parasites and gain immunity; the parasites developed methods to overcome the immunity; and the cycle would repeat. Eventually, organisms evolved that seemed to be potential victims but that actually had mechanisms allowing them to fool the parasites and cause them to clone the victim rather than themselves!

We now describe the Avida system in more detail and discuss how it can be used as a system for investigating evolutionary processes. An Avida run starts by the execution on the virtual machine of an initial organism that is capable of self-replication. This organism is the initial input provided by the user. For example, the initial organism described in Table 6.3 is capable of self-replication. The program is written in the machine language implemented by the Avida system. There is no need to try to understand the details of the machine language, as we just want to give a general notion of what the digital organisms look like. What follows is a high-level description of the organism's operation.

The program starts out by allocating memory for the future descendant. Then the program seeks its end, which is marked by the two commands **nop-A** ("nop" stands for no operation) and **nop-B** (these commands do not do anything). The new copy will be written to this memory location, where the newly allocated memory resides. Note that the template **nop-A**, **nop-B** is represented for the **h-search** command by the template **nop-C**, **nop-A**, which appears in the next two lines of code (this is how the machine language of Avida uses **nop** operations to represent labels). After this initial step the copy loop starts executing and will execute as long as the template **nop-A**, **nop-B** (represented again for the **if-label** as **nop-C**, **nop-**A), which marks the end of the program, has not been copied. For every loop cycle one command is copied (by **h-copy**) from the *read head* (which is at the beginning of the code at

TABLE 6.3   A Description of an Organism in the Avida System

# — Setup —	
h-alloc	# Allocate extra space at the end of the genome to copy the offspring into.
h-search	# Locate an A:B template (at the end of the organism) and place the Flow-Head after it.
nop-C	#
nop-A	#
mov-head	# Place the Write-Head at the Flow-Head (which is at beginning of offspring-to-be).
nop-C	# [Extra nop-C commands can be placed here w/o harming the organism!]
# — Copy Loop —	
h-search	# No template, so place the Flow-Head on the next line (to mark the beginning of the copy loop).
h-copy	# Copy a single instruction from the read head to the write head (and advance both heads!)
if-label	# Execute the line following this template only if we have just copied an A:B template.
nop-C	#
nop-A	#
h-divide	# ...Divide off offspring! (note if-statement above!)
mov-head	# Otherwise, move the instruction pointer (IP) back to the Flow-Head at the beginning of the copy loop.
nop-A	# End label.
nop-B	# End label.

Source: Courtesy of Charles Ofria.

the beginning of execution) to the *write head* (which the program positioned at after the template **nop-A, nop-B**). At the end of the loop (after copying the whole program), the **h-divide** command is executed and causes the program in the new memory region to turn into an independent organism. The experimenter using Avida defines the probability that **h-copy** will misbehave, and, instead of copying the command it is supposed to copy, a random command will be copied into the descendant. This creates mutations.

Note how Avida's machine language uses the **nop** commands, whose execution has no effect, to seek memory locations (**h-search**) and to check which commands have been copied (**if-label**). It is worth reflecting on how this design influences the ease of writing self-cloning programs compared with addressing memory in the standard ways used in other machine languages.

The Avida developers mention a few factors demonstrating the appeal of digital organisms for evolutionary research. Here are two of the most important ones:

- Researching artificial life allows us to generalize about systems with self-replication capabilities. This allows us to study other evolutionary systems in addition to the biological ones based on DNA and RNA.

- Studying digital organisms allows us to discuss questions that cannot be researched using biological systems. For instance, a certain type of mutations can be canceled, or certain evolutionary stages can be pinpointed so that changes can be made to them.

Among the topics studied using Avida are the importance of the fitness of intermediate evolutionary steps in the evolution of complex properties, the factors causing the coexistence of multiple species rather than the creation of a single dominant species, and the evolution of cooperation.

Clune, Ofria, and Pennock (2007) studied the evolution of plasticity and made an interesting use of Avida (see Chapter 3 for a further discussion of plasticity). They tried to test whether frequent changes in the demands made by the environment will cause the digital organisms to evolve an ability to adapt to the changing environment, that is, to evolve behavioral plasticity. Recall that in Avida the organisms may be required to fulfill some tasks, and success at the tasks gives them additional running time (failure may result in a decrease in the allotted running time). The researchers added a mechanism allowing the organism to determine if the task was accomplished successfully (using the return value from the output command). The aim was to test if this input from the environment will be used by the organisms to adapt their behavior to the demands of the environment, which was changed cyclically between requirements for two different computations on the input values. The researchers hoped that the system will create organisms which check the return value of the output command and use it (using a conditional statement) to choose a computational path providing for more running time. Dramatically, something else happened: the artificial evolution managed to find an organism with one computation path that did not use the return value in any conditional statement yet remained adapted to the changing environment. The organism achieved this by using the return value as one of the numbers used in the computation of the next output value, in a way that assured the new output generated matched the environmental requirements represented by either one of the two return values. In other words, evolution managed to create an organism that did not need plasticity at the level of the computation paths (while the output it produced obviously had to demonstrate the required plasticity). To force the system to develop organisms with developmental plasticity of the type they were hoping to find, the researchers had to avoid the creation of such a static solutions by changing the conditions of the problem in a subtle way aimed at undermining solutions with a single computation path. The variety of environmental demands in nature may not allow for the creation of "sneaky" solutions such as those initially created.

From a scientific perspective it is important to note that similar results about the evolution of plasticity and its relation to environmental demands were also obtained in studies looking at the evolution of neural networks under changing environmental conditions. The fact that similar evolutionary phenomena arose in models that are fundamentally different from each other allowed the researchers to draw from the results of the different experiments general conclusions about the behavior of evolutionary processes.

6.3.2 Evolvable Virtual Creatures

Karl Sims (1994) presented a way to use ideas derived from artificial life to create three-dimensional graphical creatures for animated movies. These creatures had to be mobile in their environment and to react to external stimuli. Their shape and the "brain" controlling their musculature and reacting to the environment are generated in the Sims system by a genetic algorithm of the sort described in Chapter 3. The evolution of the creatures can be directed toward certain behaviors by choosing different fitness functions. Sims presented a large collection of creatures that have evolved in the system and were able to walk, swim, jump, and follow light sources.

The genotype of the creatures determined both their physical structure and their control mechanism. The control mechanism was not a standard neural net as described in Chapter 4 but rather contained more complex elements capable of performing operations such as addition, multiplication, conditional statements, and mathematical functions such as *sin, log,* and *abs.* The genotype was a graph representing the rules for building the creature (as can be seen in Figure 6.3). The graph contains both the representation of the creature's physical structure and the description of the neural net and sensors allowing it to sense its environment. As seen in Figure 6.3, a creature is constructed from a collection of connected blocks.

Genotype: Directed graph   **Phenotype:** Hierarchy of 3D parts.

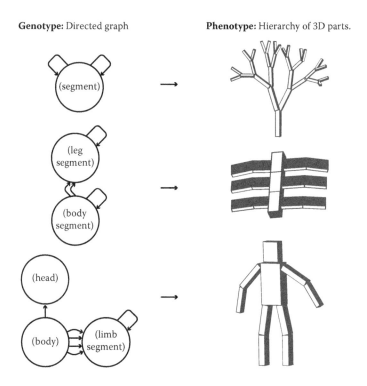

FIGURE 6.3 Sims's virtual creatures. (Adapted from Sims, Karl, in *Proceedings of the 21st Annual Conference on Computer Graphics and Interactive Techniques*, 15–22, ACM, 1994. With permission.)

Its brain reacts to environmental stimuli and determines how to move the different blocks, similar to the way a live organism moves its limbs. This movement of various body parts allows the creatures to move around its living space.

Sims performed experiments on the evolution of creatures under different conditions, such as walking on a flat terrain, swimming in water, or following a moving light source (see Figure 6.4). Fitness was computed taking the simulation's goal into account. For instance, for the evolution of walking, the fitness function took into account the distance of the creature from its starting point and its final velocity. Like in other genetic algorithms, individuals were selected for reproduction (either sexually or asexually). During the offspring creation phase, certain mutations could occur as well as chromosome crossover (in sexual reproduction).

FIGURE 6.4 Creatures adapted to walking. (Adapted from Sims, Karl, in *Proceedings of the 21st Annual Conference on Computer Graphics and Interactive Techniques,* 15–22, ACM, 1994. With permission.)

An important part of the system was the simulation of the properties of the environment: the evolution of walking and jumping was affected by gravity, whereas the evolution of swimming was affected not by gravity but by water viscosity and its influence on mobility. Moreover, the velocity of each creature was computed by taking into account the movement of its muscles and physical laws. All of this had a significant impact on the reality of the simulation. Indeed, when one watches a video of the creatures, their movement seems extremely realistic and reminiscent of the movement of animals.

Not only did Sims's simulations emulate the physical environment; the creatures existed in an environment that could contain more than one creature, and the creatures could interact with each other. For example, one of the simulations demonstrated competition between two creatures, where a creature "won" (evolutionarily) if at the end of the competition it was closer to a cube located in the environment. The creatures could look for the cube, move it, disturb each other, and so forth. Dealing with the environment in all its richness is of course a central factor in biological evolution, but many of the evolutionary computation models we have discussed thus far do not address this at all or address it only in a very limited fashion.

6.4 SYSTEMS BIOLOGY

Systems biology is a new research area attempting to apply modern techniques to the study of whole living systems rather than individual elements such as genes, proteins, or single cells. It might seem strange given how much we do not know about the basic components of biological systems to attempt to study how they combine and interact. For example, given that we do not fully understand how single genes function, how can it make sense to try to look at the next level up and study how large group of genes function together? This approach might be less ridiculous than it first sounds. In many cases, studying the interaction between components sheds light not only on the system behavior but also on the behavior of the individual components.

While not complete, the vast knowledge accumulated about the biological building blocks allows for the usage of mathematical tools and simulations to try to understand how these components interact to create complete biological systems. As we saw throughout the book, complex systems can present properties that are generated by the way the different components interact and influence each other and that do not exist at the level of the separate components.

Systems biology, then, is the attempt to construct models describing biological systems in order to investigate the interactions between the elements of biological systems, to study the behavior of these models, and to use them to explain the systemic properties of the biological systems. Many types of modeling techniques are used, and models range from mathematical models consisting of equations describing the relationships between various quantities in the system to computational models attempting to describe the step-by-step operation of the biological system being modeled (see, e.g., the discussion of statecharts later in the chapter).

An important example of a systemic property is tolerance to disturbances and "noise" (**robustness**). System robustness manifests itself in a variety of ways: adaptability to changing environmental conditions, insensitivity or low sensitivity of the system to certain changes, or gradual reaction to damage to the system rather than a catastrophic shutdown. Robustness is discussed in more detail later in this section.

A "systemic" approach to system-level properties such as robustness requires an understanding of both the structure and *organization* of a system (i.e., the components and their interconnectivity) and an understanding of its *dynamics* (i.e., how the system behaves through time and how it reacts to different conditions and forces applied to it). A systemic understanding of the organization and behavior of systems in the human body may, for example, be helpful in the development of new drugs and medical treatments that can change the behavior of these systems. It is important to note a fundamental problem in studying the dynamics of a system: the data we collect are mostly static and describe (partially) the state of the system at a given moment. Even when discovering connections between some components of the system, one still has to discover how these connections generate the observed dynamics. The mechanisms responsible for controlling a system's dynamics are called **control mechanisms**. Reasoning about their behavior based on observations is not at all simple and makes use of, for example, statistical tools and simulations. This analysis can be likened to **reverse engineering:** the attempt to understand the operation of a technological system such as a computer program or complex machine by observing its behavior. The goal is to be able to describe the biological system's behavior in a precise quantitative manner so that we can analyze it analytically or simulate it and thereby discover answers to questions about the system, some of which may be very expensive or even impossible to study directly on the system itself. For instance, using simulations of biological processes in drug design can reduce the need for animal experimentation and can shorten the development cycle, thereby reducing the development cost while also allowing the researchers to test how the drug will behave in rare conditions. For the simulation to be useful, it has to be as precise as possible and take as many factors as possible into account. All of these difficulties are examples of the challenges facing systems biologists.

As mentioned in Chapter 1, biological systems present a wide array of hierarchical organization levels—starting with organic molecules such as DNA and RNA, moving on to the organelles that build up the cells, then the cell, which is a basic unit capable of surviving and reproducing independently, and then on, in multicellular organisms, to tissues, organs, and the whole multicellular individual. Multiple individuals create populations and communities that are complex dynamical systems.

Systems biology builds models for different levels of organization. Some models focus on whole subsystems (e.g., the processes responsible for managing blood sugar), whereas others focus on one process built up from a few interconnected stages. As of now, most models limit themselves to subsystems. There is an interesting and important project to simulate all the molecular life processes in a minimal cell (i.e., a nonspecialized cell). The goal of the project is to present in a quantitative and exact way the set of basic life processes necessary to maintain a cell (see http://www.e-cell.org for more information). Another kind of systems biology model being developed is the **whole patient model**, which attempts to simulate a patient for drug development purposes (e.g., *Entelos*® provides a technology called "Virtual Patients"). Such models have to address the different organizational levels of the patient (e.g., the links among genes, chemical processes, intercellular communication, and the organization of tissues, organs, and finally the whole patient). Each structural level may operate at different scales of size and time rates and present different types of behavior. To be true to life and useful, a model has to capture the interactions of the different organization levels.

One of the research goals is to identify and characterize *modules* or biological "circuits" with well-defined roles that are used as building blocks in the assembly of the more complex biological systems, similar to the way electronic circuits are used to build computer systems. Researchers have been successful in identifying the control mechanisms that determine the properties of many such modules. The modules contain proteins that act together as an organized system with a well-defined goal or are made up of cooperating gene (or protein) networks. Examples of such modules and their control mechanisms include positive feedback loops, negative feedback loops with delay mechanisms, mechanisms that implement temporary storage of data (memory), noise-reducing and noise utilization mechanisms, and various oscillators (Kitano, 2002). Identifying the biological and chemical ways these control mechanisms must be implemented to create the required behavior allows us to understand the biological systems at a high level of abstraction based on engineering descriptions of the characteristics common to different processes sharing the same control mechanisms. The engineering approach allows us to use the same mathematical tools used in system engineering (e.g., differential equations). Obviously, the same control mechanism may be implemented in several ways at the chemical level; nonetheless, understanding the control mechanisms and the ways the different modules interconnect to create a whole biological system gives us a new perspective on biological systems.

It is interesting to note that, in contrast to engineered control mechanisms, which are designed to implement desired behavior, the biological control mechanisms evolved as a result of various evolutionary needs over long periods of time. It is natural to wonder about the chances that evolution will give rise to modules that exactly match the control mechanisms developed by engineering disciplines. To answer this question, different biological control mechanisms have to be identified and analyzed, and their evolution must be investigated. There is no doubt that our ability to perform large-scale studies and to analyze data from many sources presents large challenges to systems biology and high expectations for new biological insights. Time will tell whether computational systems biology is up to this challenge.

We now discuss two examples of questions asked by systems biology: (1) the origin and nature of biological modularity; and (2) the robust architecture of gene networks. We conclude this section with a discussion of the application of formal languages to the description of biological systems.

6.4.1 Evolution of Modularity

The modular and hierarchical structure of organisms (which contain cells, tissues, and organs) raises the obvious question about the evolutionary benefits of such a structure, and this question is the focal point of many studies. **Herbert Simon**, one of the central figures of artificial intelligence during its heyday, offered one famous explanation in his paper titled "The Architecture of Complexity" (1962). He defined the term **nearly completely decomposable system (ND)** to describe systems made up of separate components in which there is much more interaction within each component than between different components. It is easy to see that many biological and physical systems fall into this category. ND is not the same as modularity, as can be seen from the variety of properties of the previously mentioned biological "circuits," but it does define a central property of modular systems. So the question is how evolution leads to ND systems. Simon answered this using a parable about two watchmakers, named Hora and Tempus. The watchmakers build almost identical watches, each of which contains 1000 components. The difference is that Hora builds his watches out of 10 stable modules, each containing 10 stable submodules with 10 elements each. Tempus, on the other hand, does not use such stable substructures, and the only stable structure he comes up with is the whole watch made up of 1000 pieces. Assume both watchmakers are distracted frequently by phone calls from their customers. Clearly, Hora, who has to assemble only 10 modules between interruptions, will be much more productive than Tempus, who needs an uninterrupted period of time long enough to assemble 1000 elements and has to restart from the beginning after each phone call. Hora, Simon tells us, prospered, while Tempus grew poorer and poorer and finally lost his shop. While Simon's parable is told about watchmakers, it is in fact concerned with the organization of the watches, as examples of two ways complex systems might be organized, rather than with the role of the watchmakers or the origin of the different organizations.

Simon concluded from the parable that in an evolutionary scenario ND watches (e.g., Hora's watches), or in general ND systems, would be fitter than their non-ND counterparts and therefore will have the upper hand in the evolutionary race (the reader is encouraged to understand how the conclusion about fitness arises from the parable).

As already noted and acknowledged by Simon, this model is very general and allows for different conclusions about the evolutionary process (note, in particular, that the model does not address the evolution of modularity of watches per se, but only deals with its advantages if it exists). Simon also has a stronger claim about ND systems: such systems will improve their fitness faster than non-ND systems with similar complexity (the property of how well a system can undergo evolutionary changes is called its **evolvability**). The reason is that an ND system allows for local changes and therefore raises the probability that a change in one of the components improving fitness will not compromise other components (Simon, 2002).

The watch discussion seems to have dealt with the phenotype of a system, but the evolvability claim actually suggests that the genome may be ND in some sense and is reminiscent of Holland's building block hypothesis (see Chapter 3). Biological systems present both genotypic and phenotypic modularities. For instance, each of our two arms is a defined organ, and an arm injury does not directly affect other organs; therefore, in this sense each arm is a module. On the other hand, both arms reflect the same genetic template and not two different genetic modules. However, we did see in the previous discussion of biological circuits that we can identify sets of genes operating as separate modules in an ND-like fashion. Simon uses the notion of ND to discuss both these aspects of modularities, which do not necessarily arise due to the same reasons and the same evolutionary pressures. ND is useful in discussing modularity but does not explain the difference between these two kinds of modularity.

6.4.2 Robustness of Biological Systems

An important property of living organisms is their robustness to various internal or external mishaps occurring before and during their lifetime, including genetic mutations, developmental perturbations, and accidental events. We will not define robustness here, and it is clear that too many accidents will cause an organism to fail and eventually to die; however, our daily experience convinces us that organisms are generally robust to many such events.

Engineering has taught us a variety of methods for achieving robustness. The central ones are as follows (Kitano, 2002):

- **Control mechanisms***: In particular, negative feedback.

- **Modularity***: Allows for the containment of failures so that a failure affecting one module will not spread and cause a total system failure.

- **Redundancy**: A few components with identical functions can serve as backup for each other.

- **Structural stability:** A physical structure can provide stability.

These methods are also available to biological systems. A simple example is the redundancy achieved by having many different cells with identical functionality (consider the huge number of blood cells, which is the reason a minor injury does not cause a significant physiological problem). The engineering knowledge of these methods for achieving robustness can help with the understanding of biological processes or at least can aid in creating exact mathematical models that will be the basis for new research.

Robustness of Gene Networks

One can perform large-scale experiments on simple organisms such as bacteria, worms, and yeasts where a single gene is removed from the genome or deactivated and the effect on the phenotype is studied. For instance, it turns out that 82% of the 6000 genes of yeast are not strictly necessary, and removing each of them leaves viable strains (Giaever et al., 2002). Moreover, only 15% of these genes affected the organisms' rate of growth. In other words, 70% of the yeast's genes do not seem to adversely affect its functions when missing. Clearly, we cannot judge the quality of life of these organisms, and probably this research has to be repeated under various environmental conditions where the affects of the loss of the genes may be more pronounced. In any case it is clear that a very large fraction of the yeast's genes are not absolutely necessary. Similar results were obtained when removing a single gene from organisms of varied complexity, from bacteria to mice.

It would seem that these results suggest a simple mechanism that can explain the robustness of organisms. Genes may have backups that become operational when their counterpart is missing; therefore, removing a single gene at a time is not likely to cause any harm. This is similar to the engineering practice of increasing robustness by redundancy, for instance, by having dual wheel retraction systems in passenger jets.

However, this simplistic explanation raises two issues. The first is that evolution can almost never preserve a gene whose sole purpose is to protect against mutations. This is because mutations are rare events, so there is no obvious advantage in having a redundant gene; as a result the backup gene will accumulate mutations over time and eventually lose the ability to produce the backup protein. The second problem is due to results of recent large-scale studies where pairs of genes were deactivated. If deactivating each of the genes separately would not seem to affect the system but mutual gene-pair deactivation had catastrophic results, then this could indicate that the two genes act as backups for each other. The results of the experiments present a much more complex picture (Tong et al., 2004). Very few pairs that back each other up have been identified, and most genes are members of modules interacting to create complex webs of functional modules.

As a result, the current view is of partial backup among genes (or modules) that have some functional overlap. In this way, each gene has a specific role that affects the fitness of the organism; thus, it would be valued and preserved by evolution. On the other hand, the gene can at least partially substitute another gene if that gene fails (Kafri et al., 2005). For example, there might be two enzymes that digest different types of sugars and can stand in for each other in a partial fashion (e.g., in lesser efficiency) in case one enzyme fails.

6.4.3 Formal Languages for Describing Biological Systems

Research in systems biology uses advanced computational capabilities to build simulations and to test hypotheses using models. To be able to describe the models in a consistent and uniform way and to share data between different research groups, formal languages for describing the biological models were developed. Currently, biological information is described (e.g., in textbooks and most scientific papers) in natural language accompanied by figures and pictures, but these descriptions can be ambiguous and unclear. Hence, the need for formal languages with precise semantics to describe biological models. Such languages are very important when we want to make sure that information collected by multiple research groups is consistent. In addition, they can be used as input languages for programs that simulate a model and describe its behavior graphically and as output of programs for visual building of models. Another important usage of formal languages is the ability to publish in conjunction with a traditional scientific paper an exact description of the model in a standardized language allowing other researchers to evaluate the results more easily. To these ends a few languages based on Extensible Markup Language (XML) have been defined. The best-known among those languages is **Systems Biology Markup Language** (**SBML**). Using SBML one can describe biochemical networks, that is, systems composed of a collection of chemical objects (e.g., molecules) linked to each other by chemical reactions. Using standard languages such as SBML allows data repositories containing a large collection of models from the scientific literature to be created making them available to the research community.

One can also go a step further and use tools developed for modeling computer systems to build "active" biological models. An example of this is the use of **statecharts** to build formal models of biological systems. The statecharts language is a visual language developed in 1984 by **David Harel** to aid in developing complex **reactive systems** (this language was originally intended to be used in the development of aeronautical systems). The behavior of a system is described using states and events that cause transition between states. The states in statecharts may be composed from substates, allowing the specification of systems at different levels of organization, and for easy transitions between levels of description. Moreover, using statecharts one can allocate states to components acting in parallel and thereby can describe systems containing parallel and interacting processes. In contrast to a verbal description, a system described by a statechart is defined exactly and therefore allows for automatic execution.

Researchers have used statecharts to describe different biological mechanisms, including major processes of the immune system. These models were used to integrate the data obtained from many decades of research and to test by simulation whether these data are consistent and whether the models agree with the observed behavior of the immune system. To understand a system described using statecharts, one can use the simulation tools originally developed to interact with models of computer systems. These tools allow viewing animations that visualize the behavior of the system in order to observe the state of each object during the execution, and to change objects' states. All of these are, of course, necessary to verify a model and to understand its behavior.

Figure 6.5 (from Setty et al., 2008) is a statechart of a eukaryotic cell in a multicellular organism, which is specified by using three distinct objects, namely, the nucleus, membrane, and cell. The cell includes the specification of the different stages in the life cycle of the cell. The nucleus object specifies gene expression in a discrete fashion, whereas the membrane object specifies the response to external stimulations. The statechart of the cell object contains two concurrent components: the proliferation and differentiation processes. The proliferation component defines a state for each stage of the cell cycle, whereas the differentiation component specifies

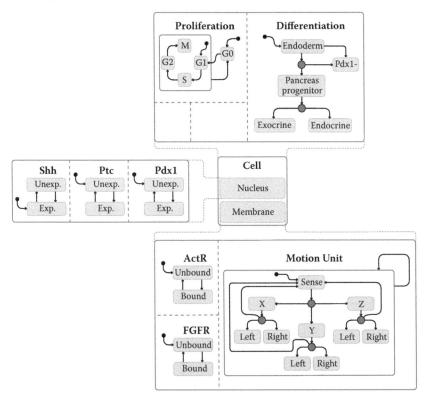

FIGURE 6.5   Statechart description of a eukaryotic cell. (Adapted from Setty, Yaki, Irun R. Cohen, Yuval Dor, and David Harel, *Proceedings of the National Academy of Sciences* 105, no. 51, 20374–20379, 2008. With permission.)

a state for each developmental stage of the organism. The nucleus and the membrane objects are located inside the cell to indicate the (strong) composition relation among the three objects, that is, that the nucleus and the membrane cannot exist without the cell containing them. The statechart for the nucleus specifies each gene as an independent component that can be either in an expressed or an unexpressed state (denoted by Exp. and Unexp., respectively). Three genes—Sonic hedgehog (Shh), Patched (Ptc), and Pdx1—involved in pancreatic organogenesis are shown in this example. Similarly, the statechart for the membrane specifies the cell's reactions to possible external stimulations. Two subcomponents within the membrane statechart specify two receptors in the membrane—activin receptor (AcrR) and fibroblast growth factor receptor (FGFR)—that can be in a bound or an unbound state. The third component in the membrane statechart depicts the motion unit that continuously scans over six possible directions to find the optimal move. The states contain behavioral instructions for the cell. For example, in the membrane, the state bound of a receptor defines the specific genes it activates. Similarly, in the nucleus, the expressed state contains instructions for genes to activate the expression of other genes.

While we cannot go into a full description of the semantics of statecharts here, it is important to realize that this graphical representation carries a precise meaning. For example, a cell is presented from two orthogonal views (i.e., proliferation and differentiation) marked by dotted lines. In the proliferation view, the statechart tells us that a cell can be either in a resting state G_0 or in the active part of the cell cycle that must start in G_1.

The language we just described is based on *states*. One of the problems with this notation is that we often do not have enough biological data (or the data are not precise enough) to describe all the states and transitions of a complicated biological systems. Thus, other projects used languages and formal notations that are **scenario based**—for example, a list of rules that described what a cell does in a certain situation given a certain stimulus. Then, given a set of such rules, the system allows for execution of many scenarios from different initial conditions. Such languages can better cope with partial knowledge. For this reason they may be better suited at the present time for describing biological processes.

In addition to statecharts a variety of formalisms developed by computer science have been adopted for the description and analysis of biological systems. Among these are Petri nets, process calculi (e.g., the pi-calculus), and Boolean networks (Fisher and Henzinger, 2007).

The last few years have seen many initiatives for creating languages for describing biological systems. Paradoxically, this multitude of initiatives is problematic. It seems that it would be better to select one language (or a small number of languages) and to focus on the monumental task of translating the vast array of biological knowledge to this formal language. Only when there is a critical mass of biological knowledge described in a few common formal languages will we be able to gain the full scientific benefit promised by standardized formal languages.

Given the current difficulties in adapting and using a common formal language to describe biological systems, there is an ongoing effort to extract biological knowledge from natural language texts, that is, from biomedical journal articles. While regular text searches use exact word matching and keyword annotations, more sophisticated methods aim to use natural language processing (NLP) techniques combined with machine-learning algorithms using biological ontologies and dictionaries to extract knowledge from biomedical articles. While far from perfect, such text-mining systems bring hope to the endeavor of retrieving at least some of the vast amount of knowledge that has been published and converting it to machine-readable form. Having the data specified in formal languages with precise semantics will ease the goal of building systems to store, manage, and mine biological knowledge.

6.5 SUMMARY

We have outlined in this chapter a few of the varied directions taken by researchers who study computational approaches motivated by biology. Each of the topics we described is an active research area with new ideas and applications being developed constantly. The topics we presented in this chapter are newer than the "classic" areas we described in previous chapters, and it is safe to assume they will develop in varied and surprising directions.

We have attempted throughout the book to emphasize the computational aspect of nature, particularly the study of biological processes as computational processes (i.e., as information processing and problem-solving processes). This outlook made us consider which computational problems can be solved using biological mechanisms, looking at the spectrum of biological mechanisms from molecular processes to the behavior of animal populations. To this end we inspected the information the processes consume and how it is saved and manipulated; the control mechanisms responsible for various processes; the roles of parallelism and distribution; and the methods for dealing with faults, noise, and missing information. This perspective obviously does not address all the different ways of researching and studying biological processes, but it does allow us to observe aspects of the biological systems that might be obscured otherwise and to use computer science and engineering tools to help understand biological phenomena.

On the flip side, viewing biological phenomena with "computational eyeglasses" allowed computer scientists to develop new computational models inspired by biology and new methods for solving computational problems. These included optimization and search problems, clustering and classification problems, pattern recognition, and machine learning. Most of these new models are not exact representations of biological processes (which are only partially understood for the most part) but rather are new models developed by computer scientists inspired by the knowledge gained from the study of biological systems. The computational perspective that guided the discussion in this book provided insights about basic computer science ideas, including computational universality, the fundamental inability of distinguishing between programs and data, ways to build parallel and distributed systems, and dealing with and utilizing noisy data. An important property of many of the methods we discussed is that they are based on using local data and control (this is especially manifest in cellular automata, neural nets, computational immunology, and swarm intelligence). Obviously, locality is of major importance in building parallel and distributed systems.

Another recurring theme was that a system containing a large number of simple components may be much more complex than each of its components. For example, in the "Game of Life" we derived a system that is equivalent to a digital computer using very simple birth and death laws. The existence of a population of different solutions and a simple selection process allows for optimizations that cannot be achieved by a single solution. The learning and computing capacity of a neural system is much greater than that of a single neuron. Molecular computation was another example; as we saw, a large set of simple molecules can effectively solve complicated computational problems.

An important aspect of these new models is that they rely to a large extent on learning and self-organization rather than on conventional programming. Modern computer systems have to contend with more and more complex computational problems, failures of various kinds, and complex and changing environments; to adapt to input changes and sometimes even to required changes in output; and also to deal with huge amounts of data in an effort to find patterns and statistical links. These requirements are only part of the challenges faced by developers of large computer systems, and these challenges make the programming of these systems harder and harder. It is difficult to believe we will be able to avoid in the near future the need for system analysis, software design, and programming. Hopefully, we may be able to hand some of the tasks faced by the computational system to mechanisms that can deal independently with them by machine learning and self-organization.

Not only do the topics we explored in this book have a major research interest; they are also used for a wide variety of practical technological applications. The models we presented (in particular genetic algorithms and neural nets) allow us to deal with complex optimization and planning problems and with problems that involve very large amounts of data, therefore requiring huge computational resources. Using the tools we presented often helps in reducing the amount of resources needed to more manageable levels. Some of the problems do not have other feasible solutions, whereas using self-organization characteristic of biological models allows us to cope with them, either by using an evolutionary process similar to genetic algorithms or by a learning process of the kind implemented by neural networks. Examples of such problems are handwriting recognition, image recognition, and data mining.

In recent years buzzwords such as **complex systems, nonlinear systems, self-organization**, and **emergence** are often used in technological discussions to describe the behavior of dynamical distributed systems that do not employ hierarchical control. It is also common to associate properties such as learning ability, adaptability, and robustness with such systems. Often it is unclear what the exact meaning of these properties is and how to discuss them formally. We have presented in this book specific examples of systems with these properties in an effort to make them clearer and more tangible. We attempted to show how such properties manifest themselves and how they can be analyzed and made useful. We avoided theoretical definitions of these terms, and we tried to steer clear of vague generalizations. Computer scientists have dealt with the different aspects of these topics in a formal mathematical manner, and we provide suggestions for relevant further reading at the end of this chapter. In this book we preferred to focus on the diversity of biological examples while emphasizing their common properties on one hand and the richness of each biological example on the other hand.

Solving technological problems using the ideas we presented in this book often requires a combination of different methods and the use of the new models in conjunction with standard methods. For example, we can use genetic algorithms to discover a neural net with a useful topology allowing it to learn a training set efficiently and to achieve a required behavior and good generalization. Another example of a combination of several models is the simulation of a population of neural nets embedded on a grid that pass information to each other, similar to the implementation of a cellular automaton. The set of possible combinations is obviously infinite. Just as solving a new problem using standard algorithms necessitates using existing algorithms and adapting them to the new problem, the same is true when using the new methods and models.

When using the ideas presented in this book, do not hesitate to make changes to the solutions we presented. Often trial and error is the way to find successful new solutions. Sometimes a solution is possible only after preprocessing the data so it is better adapted for a particular computational model (we saw examples when discussing neural networks). Technological applications often require some changes to fine-tune the system. Do not neglect the exciting possibility of observing the world, be it the physical, chemical, biological, or human aspects of the world, borrowing ideas and turning them into computational models. There is great opportunity to develop new ideas and new applications of existing ideas!

6.6 RECOMMENDATIONS FOR ADDITIONAL READING

We recommend the following books, which deal with the topics we discussed in this book. They may be used to deepen the understanding of topics we discussed, to find more examples, and to become familiar with other computational models inspired by biology.

6.6.1 Biological Introduction

The following are some textbook suggestions for readers who are unfamiliar with basic biology or who want a deeper biological introduction than provided in Chapter 1. Many other fine textbooks are available.

Solomon, E., L. Berg, and D. Martin. 2007. *Biology*, 8th ed. Florence, KY: Thomson Brooks/Cole.
Campbell, N. and J. Reece. 2008. *Biology*, 8th ed. Pearson Education.
Starr, C., R. Taggart, and C. Evers. 2008. *Biology: The Unity and Diversity of Life*, 12th ed. Florence, KY: Thomson Brooks/Cole.

6.6.2 Personal Perspectives

Crick, F. 1988. *What Mad Pursuit: A Personal View of Scientific Discovery.*
Jackson, TN: Basic Books.
Watson, J.D. 1968. *The Double Helix: A Personal Account of the Discovery of the
Structure of DNA.* New York: Atheneum.
> The autobiographical accounts of the co-discoverers of the structure of the
> DNA molecule provide enjoyable background to some of the topics discussed
> in Chapter 1. Francis Crick was also involved in the efforts to decipher the
> genetic code, and his book provides a personal account of this research as
> well.
Hofstadter, D.R. 1979. *Gödel, Escher, Bach: An Eternal Golden Braid.* Jackson, TN:
Basic Books.
> A personal book about the role of self-reference in explaining life and cog-
> nition. Paradoxes, music, logic, and computation are some of the themes
> woven together in this unique book. The book won a Pulitzer prize and
> enjoys what might be called a cult following.

6.6.3 Modeling Biological Systems

The following books present various approaches to mathematical modeling of biological systems.

Thompson, D.W. 1992. *On Growth and Form.* Mineola, NY: Dover.
> A famous and inspiring exploration of the living world, focusing on the role
> played by physical forces in determining the shapes of animals, this book is
> a treasure trove of magnificent examples of biology viewed mathematically.
> Originally published in 1917.
Mandelbrot, B.B. 1982. *The Fractal Geometry of Nature.* New York: W.H. Freeman.
> A detailed analysis of a large number of examples from the living and nonliv-
> ing world, by the man who invented fractal geometry.
Murray, J.D. 2002. *Mathematical Biology: I. An Introduction,* 3rd ed. Heidelberg:
Springer-Verlag.
> This book presents the classical approach of mathematical modeling of bio-
> logical phenomena using differential equations.
Nowak, M.A. 2006. *Evolutionary Dynamics: Exploring the Equations of Life.* Boston:
Harvard University Press.
> A contemporary perspective on the application of mathematical techniques
> to a variety of biological questions.
Prusinkiewicz, P. and A. Lindenmayer. 1990. *The Algorithmic Beauty of Plants.* New
York: Springer-Verlag. Available at: http://algorithmicbotany.org/papers/#abop.
> The book shows how the L-systems formalism can be used to model the growth
> patterns of plants.

6.6.4 Biological Computation

While many books have been written on the various topics we discuss, not too many integrative books tackle the emerging field of biological computation. The following books, which vary a lot in their technical level and scope, survey the entire field.

Flake, G.W. 1998. *The Computational Beauty of Nature: Computer Explorations of Fractals, Chaos, Complex Systems, and Adaptation.* Cambridge, MA: MIT Press. (See also: http://mitpress.mit.edu/books/FLAOH/cbnhtml/.)
A delightful book emphasizing the aesthetic aspects of natural phenomena of self-organizing systems.

Sipper, M. 2002. *Machine Nature: The Coming Age of Bio-Inspired Computing.* Columbus, OH: McGraw-Hill.
A well-written book describing in a nontechnical way many aspects of biological computation.

De Castro, L.N. 2006. *Fundamentals of Natural Computing: Basic Concepts, Algorithms, and Applications.* Boca Raton, FL: Chapman & Hall/CRC Press.
A comprehensive book covering many aspects of biological computation as well as other natural computational models such as quantum computation. The book can be used as reference book for many of these topics.

Floreano, D. and C. Mattiussi. 2008. *Bio-Inspired Artificial Intelligence.* Cambridge, MA: MIT Press.
A recent book covering in depth a wide range of topics related to bioinspired computing. The book is written with an engineering orientation and covers many biological systems.

6.6.5 Cellular Automata

Schiff, J.L. 2008. *Cellular Automata: A Discrete View of the World.* Hoboken, NJ: John Wiley & Sons.
A readable introduction to cellular automata and their applications.

Wolfram, S. 2002. *A New Kind of Science.* Champaign, IL: Wolfram Media.
A very ambitious book trying to demonstrate in detail (the book contains 1197 pages) that the entire universe around us (e.g., biological, physical, and computational phenomena) could be and should be considered as cellular automata. While the approach of the author is a matter of heated discussion, the book is thought-provoking and contains many interesting examples.

Ilachinski, A. 2001. *Cellular Automata: A Discrete Universe.* Hackensack, NJ: World Scientific Publishing.
A detailed (approximately 800-page) and technical exposition of cellular automata. Includes detailed discussions of various theoretical techniques for studying cellular automata behavior. The proof of the universality of Life is presented in detail. Among the topics covered are probabilistic CA, the relationship between CA and physics models, and a comparison between CA and neural networks.

6.6.6 Evolutionary Computation

Holland, J.H. 1975. *Adaptation in Natural and Artificial Systems: An Introductory Analysis with Applications to Biology, Control, and Artificial Intelligence.* Cambridge, MA: MIT Press, 1992.

Written by the father of genetic algorithms, this book is short but deep and complex and presents a wide and rich perspective on the subject.

Goldberg, D.E. 1989. *Genetic Algorithms in Search, Optimization, and Machine Learning.* Reading, MA: Addison-Wesley Professional.

A detailed and clear introduction, including code samples.

Koza, J.R. 1992. *Genetic Programming: On the Programming of Computers by Means of Natural Selection.* Cambridge, MA: MIT Press.

The classic book about genetic programming (not genetic algorithms). Describes how programs can be evolved to solve various problems.

Mitchell, M. 1998. *An Introduction to Genetic Algorithms.* Cambridge, MA: MIT Press.

A short and readable exposition of the research on genetic algorithms.

6.6.7 Neural Networks

There are many dozens of books and Internet sites dealing with all types of neural nets. Two "classic" and recommended texts are

Hertz, J.A., A.S. Krogh, and R.G. Palmer. 1991. Introduction to the Theory of Neural Computation. Reading, MA: Addison-Wesley. (See also http://www.phy.duke.edu/~palmer/HKP/.)

Comprehensive and detailed book with a mathematical focus. The book emphasizes the connection between neural nets and mathematical models derived from physics.

Haykin, S. 2008. Neural Networks and Learning Machines, 3rd ed. Upper Saddle River, NJ: Prentice Hall.

Covers many topics, written from an engineering/applicative perspective.

6.6.8 Molecular Computation

Not many books deal with this topic, and most of the information can be found in research papers, which are published at a fast rate. The following is one of the few introductory texts and presents varied models:

Calude, C. and G. Paun. 2000. *Computing with Cells and Atoms: An Introduction to Quantum, DNA and Membrane Computing.* Boca Raton, FL: CRC Press.

6.6.9 Swarm Intelligence

Dorigo, M. and T. Stützle. 2004. *Ant Colony Optimization.* Cambridge, MA: MIT Press.

The book discusses ant colony optimization (ACO) algorithms and includes chapters dedicated to ACO algorithms for the traveling salesman problem, ACO for NP-hard problems, and ACO for data network routing. Each chapter ends with a short section enumerating the main points raised in the chapter. Pseudo-code and exercises are provided.

Bonabeau, E., M. Dorigo, and G. Theraulaz. 1999. Swarm Intelligence: From Natural to Artificial System. New York: Oxford University Press.
The book presents techniques for building artificial systems derived from the analysis of social insect behavior. Each chapter focuses on a specific biological example, which is described, modeled, and from which an algorithm is then derived. Pseudo-code for each of the algorithms is provided.

Kennedy, J., R.C. Eberhart with Y. Shi. 2001. Swarm Intelligence. San Francisco, CA: Morgan Kaufmann.
A readable and thorough discussion of swarm and collective intelligence, as well as related areas. Includes detailed discussions of applications and philosophical and theoretical implications.

6.6.10 Systems Biology

Alon, U. 2006. *An Introduction to Systems Biology: Design Principles of Biological Circuits.* Boca Raton, FL: Chapman & Hall/CRC Press.
A recent book by one of the pioneers of the field. The book concentrates on several well-known biological examples and demonstrates how mathematical treatment can be used to gain insight into the way biological systems work.

6.6.11 Bioinformatics

In the last few years many books dealing with bioinformatics from every possible aspect were published.

Mount, D.W. 2004. *Bioinformatics: Sequence and Genome Analysis,* 2d ed. Cold Spring Harbor, NJ: Cold Spring Harbor Press.
This frequently used book goes beyond listing bioinformatic tools and databases and tries to explain, often in detail, the computational and biological background for the main tools developed in the field. The book is used as a textbook in many bioinformatics courses but is also suitable for self-study.

Lesk, A. 2002. *Introduction to Bioinformatics.* New York: Oxford University Press.
Another comprehensive book by one of the founders of the field. The book gives special emphasis to the structural aspects of biological molecules.

6.7 FURTHER READING

Bonabeau, Eric, Marco Dorigo, and Guy Theraulaz. 1999. *Swarm Intelligence: From Natural to Artificial System.* Oxford: Oxford University Press.

Bonabeau, Eric, Marco Dorigo, and Guy Theraulaz. 2000. Inspiration for optimization from social insect behaviour. *Nature* 406, no. 6791, 39–42.

Clune, Jeff, Charles Ofria, and Robert Pennock. 2007. Investigating the emergence of phenotypic plasticity in evolving digital organisms. In F. Almeida e Costa, L. Rocha, E. Costa, I. Harvey, and A. Coutinho (Eds.), *Advances in Artificial Life*, 74–83. New York: Springer.

Fisher, Jasmin and Thomas A. Henzinger. 2007. Executable cell biology. *Nature Biotechnology* 25, no. 11, 1239–1249.

Giaever, Guri, Angela M. Chu, Li Ni, Carla Connelly, Linda Riles, Steeve Veronneau, et al. 2002. Functional profiling of the Saccharomyces cerevisiae genome. *Nature* 418, no. 6896, 387–391.

Hofmeyr, Steven A. and Stephanie A. Forrest. 2000. Architecture for an Artificial Immune System. *Evolutionary Computation* 8, no. 4, 443–473.

Kafri, Ran, Arren Bar-Even, and Yitzhak Pilpel. 2005. Transcription conrol reprogramming in genetic backup circuits. *Nat. Genet.* 37, no. 3 (March), 295–299.

Kennedy, James, Russell C. Eberhart, with Yuhui Shi. 2001. *Swarm Intelligence*. San Francisco: Morgan Kaufmann.

Kitano, Hiroaki. 2002. Computational systems biology. *Nature* 420, no. 6912, 206–210.

Lenski, Richard E., Charles Ofria, Robert T. Pennock, and Christoph Adami. 2003. The evolutionary origin of complex features. *Nature* 423, no. 6936, 139–144.

Ray, Thomas S. 1992. Evolution, ecology and optimization of digital organisms. Santa Fe Institute working paper 92-08-042. Available at: http://life.ou.edu/pubs/tierra/.

Setty, Yaki, Irun R. Cohen, Yuval Dor, and David Harel. 2008. Four-dimensional realistic modeling of pancreatic organogenesis. *Proceedings of the National Academy of Sciences* 105, no. 51, 20374–20379.

Simon, Herbert A. 1962. The Architecture of Complexity. In *Proceedings of the American Philosophical Society* 106, 467–482.

Simon, Herbert A. 2002. Near decomposability and the speed of evolution. *Industrial and Corporate Change* 11, 587–599.

Sims, Karl. 1994. Evolving virtual creatures. In *Proceedings of the 21st Annual Conference on Computer Graphics and Interactive Techniques,* 15–22. ACM.

Tong, A. H., Lesage, G., Bader, G. D., Ding, H., Xu, H., Xin, X., Young, J., Berriz, G. F., Brost, R. L., Chang, M. et al. 2004. Global mapping of the yeast genetic interaction network. *Science* 303, no. 5659, 808–813.

6.8 EXERCISES

6.8.1 Swarm Intelligence

1. Is the pheromones mechanism based on positive or negative feedback?

2. To determine how ants discover paths to a food source, a device similar to the one in Figure 6.6 was created. The ants walk along the circular corridor. The bottom vertical line defines the ants' starting point and the top vertical line the location of the food source. The ants start out as depicted on the left. Initially, they distribute themselves uniformly between the short and long paths as can be seen in

the middle. After a while they concentrate on the short path as seen on the right. Try to explain the development of this organization.

FIGURE 6.6

3. How will increasing the number of ants in the ants colony optimization affect the results of the algorithm?

4. Raising the rate of evaporation of pheromones (i.e., the decrease in the amount of pheromones at every iteration) can help avoid a too rapid convergence into a suboptimal region of the search space and can also help avoid premature convergence. Why?

5. Developers who built a PSO-based system discovered that if initially ω is large (close to 1) and then is decreased after a few iterations to $\omega \approx 0.5$, the system performs better. Try to explain this observation.

6. Why is *rand()* used in the PSO rate update formula?

7. In a well-known variation of PSO, the particles are not affected by all the other particles in the population but only by a subset of "neighbors" (the set of neighbors for each particle is defined ahead of time).

   a. What changes to the PSO algorithm we presented have to be made to implement this mechanism?

   b. Discuss possible advantages of this mechanism compared with the standard PSO algorithm.

6.8.2 Artificial Immune Systems

8. Immunological detectors that react to the organism are dangerous, but on the other hand detectors that do not react at all are useless. How would you balance between these two requirements when using a negative selection algorithm?

9. How will increasing the activation threshold τ and the training period impact the frequency of false alarms?

10. Why is it important for the match counter to decay over time?

11. Ideally, the immune system should not react at all in any situation that does not pose a real risk. If that is impossible, one would want the severity of the reaction to be proportional to the probability that the situation poses a real risk. Suggest ways to achieve this in an artificial immune system.

12. Discuss the possible disadvantages of the co-stimulation method used by LISYS.

13. LISYS also uses another mechanism (not described in the text) that allows each computer on the network to have a different sensitivity level w_i ($i = 1,2,..,n$) to immunological events. The local activation threshold for the detectors in computer i is defined to be $τ - w_i$ (the higher the local sensitivity, the lower the required activation threshold). Whenever a detector's counter changes from 0 to 1, the relevant w_i increases by 1. Just like the other counters, w_i decays with time. What role does the local sensitivity mechanism play? How does it contribute to the behavior of the system?

14. The *self* set may change over time (e.g., the normal communication patterns in the network may change over time). How can an immune system of the kind described deal with this? A good solution will minimize both the number of false positives and false negatives.

15. Another mechanism used by the immune system and not discussed so far is rapid "evolution" of detectors, based on their similarity to a suspicious element. The only detectors that partake in this process identify the suspicious element with high enough confidence. They are quickly cloned, allowing for a high mutation rate. The mutation rate for each detector depends on how well it detected the suspect element—the more closely it matches the intruder, the lower its mutation rate. How can we integrate this mechanism in the detection system, and what are its possible contributions to the system?

16. The immune system naturally has to be robust. Identify the properties of artificial immune systems that contribute to their robustness.

17. We have mentioned that the biological immune system may have regulatory functions. Try to suggest possible regulatory functions for the immune system, and assess their likelihood (keep in mind the observation that the distinction between self and nonself may be the result rather than the reason for the existence of such functions). Suggest observations or experiments to test your hypotheses.

6.8.3 Artificial Life

18. Try to find a counterexample for each of the properties used to define life that were mentioned in the text; that is, describe a system that has that property but that is not considered to be alive.

19. Check to see whether the dictionary definition of the word *life* can serve as the definition of life.

20. Which of the properties we enumerated to define life presents the most difficulties to the proponents of strong ALife?

21. Both Avida and genetic programming (see Chapter 3) deal with evolving computer programs. What are the fundamental differences between the two approaches?

22. Download Avida's source code from the Internet, and study the **Divide_DoMutations** subroutine (which is part of the implementation of Hardware_CPU). This subroutine is executed after the organism executes the **h-divide** function. For which types of mutations is this subroutine responsible?

23. In Avida the basic cloning mechanism (i.e., command copy) is implemented by the virtual computer, as opposed to biological systems in which the replication mechanisms are part of the organism. Can organisms that are able to correct copying errors exist in Avida? The basic copying mechanism in Avida cannot improve evolutionarily since it is not part of the organism. Suggest a way to overcome this limitation to make Avida more faithful to biological systems. Why do you think the Avida developers preferred not to implement such a mechanism?

24. Which of the two approaches to fitness—Avida's or Sims's virtual creatures simulation—is closer to the meaning of fitness in natural selection?

6.8.4 Systems Biology

25. Find more examples of robustness, and determine whether the robustness stems from positive feedback, negative feedback, or some other mechanism.

26. Explain the impact that understanding of control mechanisms may have on applications such as drug design and development.

27. Suggest situations in which the mechanisms giving the organisms their robustness helps disease processes in becoming robust and hard to treat.

28. A heating system controlled by a thermostat operates by comparing the room temperature to the target temperature and adjusting the heating level accordingly. The obvious goal is for temperature to reach a steady state. Describe in detail the system's algorithm. Does the system implement positive or negative feedback?

6.8.5 Programming Exercises

29. Implement a solution to the traveling salesman problem using ACO. Test on a variety of graphs. For which graphs does the system fail to find a good solution? Try to add mechanisms that will improve the behavior of the system in these cases. Try to minimize the number of additional mechanisms, and avoid using global information.

30. An interesting usage of clustering using ants is for plotting graphs. Let $G = (V,E)$ be an undirected graph. The goal is to embed the nodes in the Euclidean plane such that the connections are as clearly drawn as possible. In particular:

   a. Clustered nodes will be placed close together on the plane.

   b. The distances inside clusters will be minimal.

   c. Different clusters will be far enough away from each other.

   Use the clustering algorithm we presented, in which the distance between two nodes is defined to be

$$d(v_i, v_j) = \frac{|D(\rho(v_i), \rho(v_j))|}{|\rho(v_i)| + |\rho(v_j)|}$$

where $\rho(v_i)$ is the set of neighboring nodes of v_i (including v_i itself) D is the symmetrical difference between the sets, and $|...|$ denotes the number of elements in the set. Test this solution on a few graphs (observe how the nodes move as the algorithm progresses). Has the algorithm found presentations that seem visually successful for the graphs you tested?

31. How can one use PSO to discover weights for a multilayered neural net with full connectivity, assuming a training set composed of (input, output) pairs is known? Try to use this method for one of the neural nets described in Chapter 5 (without learning), and analyze the quality of the PSO algorithm's results for varying parameters.

32. Design an artificial immune system that will alert about unusual data in a software system's output files. Assume that the software system runs a complicated computation in batch mode every night (e.g., a payroll). The immune system will be used to inspect the output files and will alert if the output differs enough from the norm to indicate a possible error (or deliberate attack) in the system. Note: the goal is to be able to identify suspicious result, not merely corrupt output files.

33. Install Avida, and program a new self-replicating organism. (Use Avida's documentation for details of the machine language.)

6.9 ANSWERS TO SELECTED EXERCISES

2. Initially there are no pheromones, and the ants choose a random path. As the left path is shorter, the ants traversing it will reach the food quicker, and when they look for a path back to the nest they will go back on the same path, following the pheromone trail they laid down. (As the other ants have not reached the food source yet, there is no pheromone trail on it close to the food source.) This choice makes the left path even more attractive, and more ants will choose it in the future.

3. Adding ants allows for better scanning of the graph and therefore may improve the algorithm's results. On the other hand, too many ants will cause the entire graph to be covered with pheromones, which will obfuscate the paths and will impact the algorithm's results negatively.

9. Increasing τ and T (up to reasonable values) lowers the rate of false positives.

17. Inflammation is a complex biological process involving the immune system. It was suggested that immune systems have a regulatory role in the inflammatory process.

21. The most significant difference is that in Avida the organism is responsible for self-replication and that no external selection mechanism controls the process. A related difference is that in genetic programming one has to define a fitness function a priori to evaluate the different solutions, while in Avida the evaluation is based only on the individuals' self-replication ability, though it is important to note that in Avida one can reward individuals for performing external tasks. Another difference is in the representation of individuals: in genetic programming the individuals are usually represented as expression trees to allow an easy crossover between the genotypes of different individuals (sexual reproduction). In Avida the individuals are represented as computer programs written in the Avida machine language.

26. Drugs may work by direct interference with control mechanisms; thus, understanding these mechanisms might help in designing treatment strategies. For example, if part of a process is controlled by positive feedback, one may affect the whole process using minimal intervention, such as by using a small amount of medication whose affect will be amplified by the feedback mechanism. Conversely, a process regulated by negative feedback may require large, potentially harmful doses of medication to overcome the negative feedback. In this case, finding drugs that modify the behavior of the negative feedback loop and can be administrated together with the original drug might be a solution.

27. The normal robustness processes of the healthy individual are partially responsible to the resistance of cancer cells to interferences in their reproduction and survival processes. An organism employs many negative feedback loops to adapt to changes in its environment. Cancer cells take advantage of these and other cellular mechanisms to resist the attacks the body may launch against them. In addition, anticancer drugs may become less effective because of the ability of cancer cells to adjust and become less sensitive to them. A second example

involves the lowered efficacy over time of psychiatric drugs, many of which achieve their effects by mimicking the structure of naturally occurring neurotransmitters (e.g., by attaching to the receptors of neurotransmitters and effectively blocking them). The body reacts in a variety of ways, including changes in the number and density of neurotransmitters receptors expressed, which cause the brain to readjust to the changing levels of neurotransmitters and to return over time to its activity state prior to the administration of the drug.

30. A relevant discussion and references can be found in Bonabeau et al. (1999). The discussion in this book explains why drawing graphs is a hard problem and why using clustering is not successful for all graphs.

Index

*For Product Safety Concerns and Information please contact
our EU representative GPSR@taylorandfrancis.com Taylor & Francis
Verlag GmbH, Kaufingerstraße 24, 80331 München, Germany*

T - #0015 - 160425 - C0 - 234/156/19 [21] - CB - 9781420087956 - Gloss Lamination